An Innovation Approach to Random Fields

Application of White Noise Theory

An Innovation Approach to Random Fields

Application of White Noise Theory

Takeyuki Hida
Meijo University, Japan

Si Si
Aichi Prefectural University, Japan

NEW JERSEY • LONDON • SINGAPORE • BEIJING • SHANGHAI • HONG KONG • TAIPEI • CHENNAI

Published by

World Scientific Publishing Co. Pte. Ltd.
5 Toh Tuck Link, Singapore 596224
USA office: 27 Warren Street, Suite 401-402, Hackensack, NJ 07601
UK office: 57 Shelton Street, Covent Garden, London WC2H 9HE

British Library Cataloguing-in-Publication Data
A catalogue record for this book is available from the British Library.

AN INNOVATION APPROACH TO RANDOM FIELDS
— Application of White Noise Theory

ISBN-13 978-981-238-095-1
ISBN-10 981-238-095-7

Typeset by Stallion Press
Email: enquiries@stallionpress.com

Printed in Singapore

To the memory of Fumi and Su Su

Preface

The purpose of this book is two-fold: First, we analyze random fields in line with *innovation theory*. The analysis will be carried out as an application of white noise analysis which has developed extensively in recent years. Most of the random fields to be discussed are complex systems that are parameterized by a space-time parameter t or by a manifold C like a curve and surface and are evolutionary with respect to those parameters. In what we discuss here, one can see, among the highlights of this volume, the effectiveness of the variational calculus applied to the random fields.

Our second aim is to show interesting *applications* in many areas, such as quantum dynamics, quantum field theory, statistical mechanics and molecular biology, as well as many other fields of science. There one can see significance of the innovation approach to random complex systems that are developing as the parameter varies.

The random fields that we discuss in this volume are assumed to be functionals of white noise of either Gaussian or Poisson type, parameterized by a multi-dimensional time (vector) t or by a certain manifold C that is running through a Euclidean space. Let a field be denoted by $X(t) = X(t, x)$ or by $X(C) = X(C, x)$, where x is a sample function of a white noise and t is a vector in a certain topological vector space, and where C changes within a certain family $\mathbf{C}$ of smooth, closed and convex manifolds in R^d, $d \geq 1$. We focus our attention mainly on $X(C)$, where, in particular, C is chosen so as to run through a space-time domain. Thus, $X(C)$ can be thought of as a realization of random complex phenomena varying as C deforms and being interfered with by a fluctuation, a mathematical expression of which is a white noise x.

Our main interest is the determination of the probabilistic structure of $X(C)$. There are several mathematical tools for this aim; among others, we first provide its stochastic variation, denoted by $X(C)$, obtained by an infinitesimal deformation of the C. The variation therefore expresses the infinitesimal change of the random phenomenon in question. Further if we can introduce an equation for stochastic variation, then it (or its solution) characterizes the structure of $X(C)$. The study of such variational calculus is to be a sort of stochastic calculus. It might be considered an analogue of the classical theory of calculus of variation. In fact, it is partially true, but not quite. In addition to the classical results, the stochastic part of the variational calculus plays a more important role, and its study requires some advanced theory of probability.

We shall also discuss random functions depending on a function f. There are many good examples in these cases. It is noted that some $X(C)$ discussed above can often been reduced to this case, since C may be represented by a vector-valued function.

With these assumptions, we are now ready to provide some background including the following topics which are known concepts in white noise theory:

(1) White noise with multi-dimensional parameters and with its restriction to a lower dimensional manifold. Similar facts for Poisson noise.
(2) Generalized functionals of white noise, and the S-transform that carries those functionals to non-random functionals of ordinary real valued functions.
(3) Theory of calculus of variations that provides a powerful tool in the analysis of random fields with the help of the S-transform.
(4) Infinite dimensional rotation group that describes the invariance of the white noise measure.

With this background we shall discuss **stochastic variational calculus** and its applications. We can see various applications that suggest to us new directions of our white noise analysis, in particular directions to quantum dynamics and quantum computation.

Before we come to the actual steps of the main subjects, we first need to determine the topology to be introduced in the class of random functions and stochastic processes, so that the variation is rigorously defined. The possible topologies are as follows:

(i) The mean square topology. In this case we start with a Hilbert space involving square integrable functionals with respect to the white noise measure. The norm of the space is given by the positive square root of the quadratic mean. The Gaussian fields have been extensively investigated within this framework. Given either Gaussian noise or other noise, the so called S-transform, mentioned in (2) above of random functional of noise is applied, so that we can use a modern version of the theory of functional analysis.

(ii) The convergence in probability. Since there are cases where the existence of moments of the random variables in question is not required, we want to discuss a sufficiently wide class of random functions, in particular random fields. Path-wise nonlinear operations are naturally introduced and discussed there. The use of characteristic functional is efficient in this case.

We are now in a position to speak of the most *significant characteristics* as well as the central *advantages* of our approach, although space is too limited to describe the details. They are listed below.

(1) If one is allowed to use intuitive notations of white noise and Poisson noise like $\dot{B}(t)$ and $\dot{P}(t)$, it is claimed that they are taken to be *variables* of random functions that describe the given random complex systems. As a result, we are led to introduce generalized functionals of noise. This allows us to discuss stochastic analysis for a sufficiently broad applicable area. Further, it is natural to introduce differential and integral operators in $\dot{B}(t)$ and $\dot{P}(t)$ and to establish an analysis of new kind.
(2) White noise measure enjoys rich invariance, so that the infinite dimensional rotation group can be introduced (see (4) above). This enables us to discuss a *Harmonic Infinite Dimensional Analysis.* Note that the analysis is essentially infinitely dimensional, the exact meaning of which will be illustrated in this book.
(3) It is emphasized that both $\dot{B}(t)$ and $\dot{P}(t)$ are elemental, being members of idealized innovation. With the factor dt they can serve as random measures. On the other hand, their sample functions (generalized functions) as well as functionals of them can be dealt with within our framework. Sample function-wise stochastic properties can be investigated by this idea. For detailed calculus, we introduce a wider class of those functionals, denoted by $(\mathbf{P})$, without assuming the existence of variances. Precise characters of *space-time dependency* and *causality* of evolutional phenomena can now be discussed in a natural manner.
(4) *Qualitative* probabilistic properties, like stochastic optimality and reversibility, can smoothly be discussed within our framework. In particular, one can see more clearly in the case of random fields. Further,

our setup enables us to see good connections to *quantum probability* and to *quantum dynamics.*

(5) Of course there are many applications where a person can receive a new insight and can see beautiful interplay between mathematics and other fields of science. The reader might think that this book aims mainly at some applications of mathematics, but that is not quite true. Application of mathematics only contributes little to mathematics. On the other hand, we know that applications of physics or of others to mathematics have often made vital contributions to mathematics. Part of this idea can be seen in this volume.

A final note is that most of the parts of this book are easily understood, and we may say that the book is elemental, but not elementary. However, occasionally some more difficult ideas ground this work, as the readers might guess. Thus, this book can be read on different levels, depending on the reader.

It is our regret that we did not make good connections with non-commutative geometry or some thoughts on topics related to path integrals, which can be discussed from our viewpoint. Readers will find a short note on this topic in the Epilogue at the end of this volume. Some supplementary notes related to this topic are also in the Epilogue.

T. Hida and Si Si
July 2003, Nagoya, Japan

Contents

Chapter 1

Introduction

1.1 The idea of our approach

We first describe the basic idea of white noise analysis, as a background for innovation approach to random fields. Then various motivations will come in the following sections.

Generally speaking, a guideline of the analysis of random complex system is the **reductionism**. Intuitively speaking, first we try to find a basic system of elemental random variables (sometimes, elemental stochastic process is taken) from the given system (*reduction*). Actually we wish to have the fundamental unit, in a sense, of randomness. Then, the given random system is expressed as a functional (*synthesis*) of those variables (or the process). Then, follows the *analysis* of the functional. The analytic properties of the functional can determine the probabilistic structure or the character of the given random complex system. It is noted that the random complex systems to be discussed are usually evolutional system depending on time or space-time, so that the *causality* is always involved there. Thus, the following diagram is given:

Reduction $\rightarrow$ **Synthesis** $\rightarrow$ **Analysis**

Causality is always involved.

Important cases, in fact in many cases, the system of elemental random variables is taken to be white noise, Gaussian white noise and compound Poisson noise. Gaussian white noise is often called just white noise and the second noise is called Poisson noise. Thus white noise analysis together with Poisson noise analysis will play the main role for our purpose.

With this situation in mind, it is now ready to tell the motivations of our innovation approach.

First, we consider a simple and familiar class of random complex systems. It is nothing but a class of stochastic processes. The problem of obtaining innovation of a given stochastic process $X(t)$ has been discussed by many authors by using various methods, in fact not so much related to each other and without general mathematical definition, although intuitive meaning of innovation and its roles are rather clear. A general meaning of innovation was first given in 1953 by P. Lévy in terms of stochastic infinitesimal equation, although the equation has only a formal significance. (For a rigorous definition see Section 8.1.) Following our idea of the study of random complex systems we have studied various kinds of examples of stochastic processes for which innovation can be obtained explicitly and recognized its significance in stochastic analysis.

It is noted, regarding a stochastic process, that the obtained results are satisfactory to some extent and have good applications. The innovation theory for a stochastic process would serve as a good model when we come to the class of random fields $X(C)$ parameterized by a manifold C. There is, of course, significant difference between the two cases; for $X(t)$ and for $X(C)$. As is easily understood, when the parameter varies, the variation of $X(C)$ carries quite large amount of information, compared to $X(t)$. Such an understanding is almost trivial, but in reality it is essential.

Thus, we have to give much interpretation to the importance of innovation of random field. For this purpose, the review of the study of $X(C)$ is given in a separate section.

1.2 Random functionals $X(C)$ depending on a manifold C

One may ask why a random field $X(C)$ should be investigated so extensively. To answer this question we shall show that interesting random fields will arise from the multi-dimensional parameter Lévy Brownian motion, the importance of which everybody agrees with. Here is a definition of the R^d-parameter Lévy Brownian.

Definition 1.1 A family of random variables $X(a), a \in R^d$, is called the *Lévy Brownian motion*, if it satisfies the following three conditions:

(1) It is a Gaussian system,
(2) $E(X(a)) = 0, a \in R^d$ and $X(o) = 0$, where o being the origin of R^d,
(3) $E(|X(a) - X(b)|^2) = \rho(a, b)$, where ρ is the Euclidean distance.

By definition, it is easy to prove that

$$\Gamma(a, b) = E(X(a)X(b)) = \frac{1}{2}(\rho(a, o) + \rho(b, o) - \rho(a, b)).$$

Remark 1.1 *The Lévy Brownian motion can be thought of as the most natural generalization of the ordinary Brownian motion $B(t)$. There is another generalization, called a Brownian sheet, but it is to be discussed somewhat different viewpoint from the Lévy Brownian motion.*

P. Lévy introduced the multi-dimensional parameter Brownian motion $X(a)$ in 1945 (C.R. Acad. Sci. 220), and studied systematically in his 1948 monograph [46]. Since then, there have been many approaches to the study of $X(a)$, including the results by P. Lévy himself. The $X(a)$ is a natural generalization of an ordinary Brownian motion $B(t)$, which occupies the most important position and plays the central role among stochastic processes depending on R^1-parameter. As one can recognize, the definition is given in dimension-independent fashion.

In particular, if the parameter a is restricted to a straight line passing through the origin, then it is parameterized by a real number t and we are given an ordinary Brownian motion with R^1-parameter.

Another remarkable property to be mentioned is that in studying the dependence of $X(a)$ on the parameter a it is convenient to introduce random fields that come from functions of the $X(a)$, a being restricted to some region in R^d. To illustrate this fact, some examples are provided below.

We considered the conditional expectation of the Lévy Brownian motion and an interesting fact was observed as follows.

Example 1.1 Let $X(a), a \in R^2$, be Lévy's Brownian motion, C be a smooth curve which is a loop and p be a point which lies inside of C. The conditional expectation is expressed as

$$E[X(p)|X(s) \equiv X(a(s)), a(s) \in C] = \int_C f(p,s)X(s)ds. \tag{1.2.1}$$

Namely, for the case of a circle C, the parameter is taken to be θ instead of s and then the kernel function $f(p,\theta)$ is obtained as

$$f(p,\theta) = \frac{(t^2-x^2)^2}{8t\rho(x,t,\theta-\beta)^3} + \frac{1}{2\pi}\left(1 - \frac{t+x}{2t}E\left(\frac{\pi}{2}, \frac{2\sqrt{tx}}{t+x}\right)\right),$$

where E is the elliptic function, t is the radius of the circle C, $x = |MP|$ and $\beta = \angle OMP$, in which M is the centre of C and O is the point on the circle for $\theta = 0$.

When C is taken to be an arc of the above curve C with the edge points a and b, the conditional expectation is

$$E[X(p)|X(s) \in C, a \le s \le b] = \int_a^b g(p,s)X(s)ds \tag{1.2.2}$$

in which the kernel function g is obtained as

$$g(p,s) = f(p,s) + \alpha\delta_a(s) + \beta\delta_b(s). \tag{1.2.3}$$

Example 1.2 Suppose the values of the Lévy Brownian motion on the entire line is given, p be a point which is not on the line. Then, we have

$$E[X(p)|X(s), -\infty \leq s \leq \infty] = \int_{\infty}^{\infty} f(p,s)X(s)ds, \tag{1.2.4}$$

with

$$f(p,s) = \frac{t^2 \sin^2\theta}{2\rho(s,t,\theta)}, \tag{1.2.5}$$

where

$$\rho(s,t,\theta) = (s^2 + t^2 - 2|s|t\cos\theta)^{1/2}.$$

If the Lévy Brownian motion is given on a finite interval $[a,b]$, instead of the whole line. Then, the kernel function is obtained as

$$g(p,s) = f(p,s) + \alpha\delta_a(s) + \beta\delta_b(s),$$

where

$$\alpha = \frac{t}{2a}\left(1 - \frac{t - a\cos\theta}{\rho(a,t,\theta)}\right), \qquad \beta = \frac{t}{2a}\left(1 - \frac{b - t\cos\theta}{\rho(b,t,\theta)}\right).$$

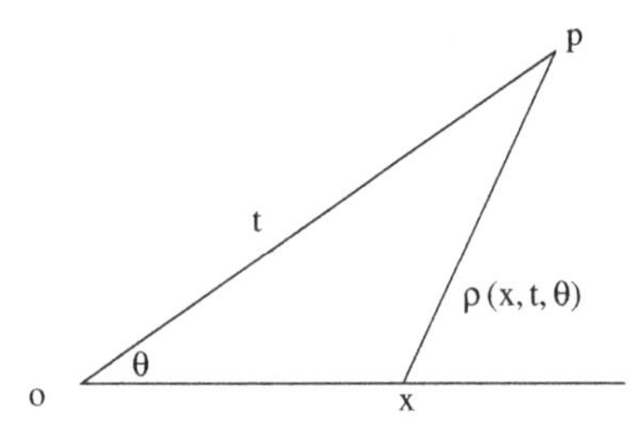

Fig. 1.

These examples tell us that when we changed from a whole loop (or entire line) to a part of the loop (or a segment), the kernel function of the conditional expectation has changes; it is a sum of the original kernel function and additional terms which are δ-functions associated with the edge points. Such an observation gives us an interpretation on how a certain singularity of the conditional expectation occurs. More generally, we can see if a manifold (taken to be a parameter) with boundary deforms, then the geometrical discontinuity of the manifold creates high singularity of the random field.

This is one of the understandings of the complex dependence on the deformation of the parameter. Thus, when the variational calculus is discussed, we assume that the manifold has no boundary and is compact.

Then, we are led to consider the variation of $Y(C)$ which denotes the conditional expectation since it can be thought of as a functional of a closed, smooth curve C. (The rigorous definition of the variation $\delta X(t)$ and $\delta Y(C)$ will be given in Chapter 7.) To emphasize the dependence of C, we introduce a notation $Y(C) = \int_C f(C,s)X(s)ds$, f: smooth.

Then, its variation $\delta Y(C)$ is obtained as

$$\delta Y(C) = \int_C \{\delta f(C,s) + \kappa f(C,s)\delta n(s)\} X(s)ds + \int_C f(C,s)\frac{\partial}{\partial n}X(s)\delta n(s)ds, \tag{1.2.6}$$

where $\frac{\partial}{\partial n}$ stands for the normal differential operator, $\delta n(s)$ denotes the distance between C and $C + \delta C$ at s, $\delta f(C,s)$ is the variation of the kernel f and κ denotes the curvature of C.

The normal derivative $\frac{\partial}{\partial n}X(s)$, is neither an ordinary nor a generalized function, however it is well defined as a generalized function over R^2 defining in a way that

$$N(\xi) = \int\int_{R^2} \frac{\partial}{\partial r}X(r,\theta)\xi(r,\theta)\,dr\,d\theta.$$

(See [70].)

This observation has made us to have interest in the variational problems of random fields $X(C); C \in \mathbf{C}$ by deforming the curve C in the family $\mathbf{C}$, which we usually take a class of smooth, convex, diffeomorphic to S^1 as a first step.

As is easily generalized, the parameter C can be taken to be a surface in a space or more higher dimensional hypersurface to discuss a random field $X(C)$. So far we took a particular and in fact important examples to know the significance of random fields parameterized by C and the roles of the variational calculus for them. Having extended such a view point, this book discusses a general class of evolutional random fields by applying the variational calculus.

(1) A general example of a random field is expressed as a *causal* representation in terms of white noise $x(u), u \in R^d$. Let $F(C, u)$ be a square integrable function of u. Set

$$X(C) = \int_{(C)} F(C, u)x(u)du. \tag{1.2.7}$$

The causality means that $X(C)$ is a function of white noise $x(u), u \in (C)$, (C) being the open or closed domain enclosed by C.

(2) By using Green's function we can form a random field $X(C, t)$, $C \in \mathbf{C}$, as

$$X(C, t) = \int_{(C)} G(C, s, t)x(s)ds, \tag{1.2.8}$$

where $G(C, s, t)$ is Green's function.

Here, it is easy to see that $\Delta_t X$, Δ_t being the Laplacian in the variable $t \in R^d$, gives the original white noise by using the classical analysis, while the variation $\delta X(C, t)$ in the variable C can be computed by using *Hadamard equation*

$$\delta G(C, s, t) = -\frac{1}{2\pi} \int_C \frac{\partial G(C, s, m)}{\partial n(m)} \frac{\partial G(C, m, t)}{\partial n(m)} \delta n(m) dm \tag{1.2.9}$$

and the innovation can be obtained.

1.3 Stochastic variational equations

Lévy's stochastic infinitesimal equation, proposed in 1953, for a stochastic process $X(t)$ is expressed in the form

$$\delta X(t) = \Phi(X(s), s \le t, Y(t), t, dt), \tag{1.3.1}$$

where $Y(t)$ is the *innovation*. Although this equation has only a formal significance, it tells us the following two important directions.

A mathematically rigorous definition of an innovation for $X(t)$ and for $X(C)$ will be given in Chapter 8.

(1) The method of constructing innovation suggests the definition of a stochastic process. After J. Bernoulli's heuristic book "Ars Conjectandi"

appeared as early as 1713, a stochastic process is understood as a random function in which a new randomness (independent of the past) appears at each instant. The randomness is expressed by the innovation. Details have been discussed in Lévy's book [47, Chapter II]. The actual formula for $\delta X(t)$ mentioned above appeared in his Berkley publication in 1953 (see [48]).

(2) The equation (1.3.1) is understood to be a formula that characterizes the stochastic process in question. It is not given to be an ordinary equation. To get a solution is not a main purpose.

What is explained above regarding the innovation is just for an intuitive interpretation and is understandable for scientists.

With this spirit in mind we shall be able to introduce a generalization of this equation to the case of a random field $X(C)$. It may be given by the following formula: when C changes from C to $C + \delta C$, we shall be given the formula

$$\delta X(C) = \Phi(X(C'), C' < C, Y(s), s \in C, C, \delta C), \tag{1.3.2}$$

where $C' < C$ means that C' is inside of C, and where $\{Y(s), s \in C\}$ is the innovation for $X(C)$. This equation may be called a *stochastic variational equation*, and it will be explained later for the innovation approach.

So far, we have given only an intuitive meaning of the equations (1.3.1) and (1.3.2) which have only a formal significance. Rigorous mathematical definition and of the innovation appeared in these equations will be given in Chapter 8.

For applicational purpose, we are much interested in the computability of calculus that are used to form the variation. In particular, that of quantum computability, so that the results would be useful to the (e.g. quantum) computer.

1.4 Random functions $X(f)$

For nonrandom functional $F(f)$, for which f runs over a certain class of functions defined on an interval $[a, b]$ with $f(a)$ and $f(b)$ are fixed, the extremal point of F is obtained by the Euler equation. The theory is well known in classical dynamics. We first see an example that will be related to randomly fluctuating functions. If f is expanded in the Fourier series by taking a complete orthonormal system, then $F(f)$ is reduced to the function with multi-dimensional parameter; in particular, R^∞-parameter case.

(1) Lagrangian dynamics. The classical mechanics can be determined by the Lagrangian denoted by $L(\dot{x}, x, t), x = x(t),\ a \leq t \leq b$. Namely, the

least action principle uniquely determines the classical trajectory $x(t)$. The action defined by

$$S(x) = \int_a^b L(\dot{x}(t), x(t), t)dt, \tag{1.4.1}$$

which is a functional of $x, \dot{x}$, and t, but eventually it is a functional of $x(\cdot)$. With this understanding we proceed to the variational calculus.

(2) Another good example is the problem of statistical hydrodynamics. We are given probability measures, depending on a time, introduced in the phase space (see E. Hopf [39] and A.S. Monin–A.M. Yaglom, vol. [58]). The evolutional equation is expressed in terms of the functional differential equation for the characteristic functional of the measure introduced on the phase space. There the functional is considered as an expectation of the evolutional random field of velocity.

(3) Since the manifold C can be expressed by a vector valued smooth function f, we may rephrase $X(C)$ by $X(f)$. Then the classical functional analysis can be applied. Note that the representation of C in terms of a function f is not unique, but the probabilistic structure is independent of the choice of f.

(4) Functionals of white noise can also be indexed by a smooth function f. Sometimes f is taken to be a kernel function. We shall see many interesting examples in applications.

(5) Multi-dimensional parameter stochastic process $X(t), t \in R^d$.

(a) Gaussian case; Functionals of (Gaussian) white noise.
The Lévy Brownian motion defined in Section 1.2 is the most important example in the multi-dimensional parameter case, and in fact it is standard from many respect.

For a random field with a multi-dimensional parameter, the multiplicity, defined in the radial direction of evolution, is usually infinite; in this sense we can see high complexity.

The dependency in the stochastic sense increases as d increases. If $d \to \infty$, a certain determinism appears. This can be seen in the case of the Lévy Brownian motion.

In addition we list the subjects to be discussed in this book, related to Gaussian systems.

(i) Markov field: The dependence on the parameter is discussed with the help of kernel functions in the representations in terms of white noise.

(ii) Reversible field (cf. 1-dimensional parameter stationary case). A generalization is obtained, where we have the invariance under certain transformations, which form a Lie group. This will be discussed in Chapter 9.
(iii) Random version of the Hadamard equation.
(iv) A randomized Lotka–Volterra equation (see Section 10.6).
(v) More on Lévy's Brownian motion.

(b) Poisson case; Functionals of Poisson noise.
(i) The theory of quantum optics, studied towards quantum computation, suggests the systematic approach to Poisson noise or that with higher dimensional parameter.
(ii) The case of the discrete chaos.
The basic element of the discrete class is given by the Wick product $:x(u_1)x(u_2)\cdots x(u_n):$ where x stands for a sample function of Poisson noise, the product is also viewed as a random measure. A field will be given by an integral with respect to this random measure.

Summing up, it is emphasized that all of them are discussed in line with white noise analysis.

To close this chapter we may say that there are three pillars upon which our white noise theory rests. They are actually advantageous of the white noise analysis.

(1) For the Gaussian case, the basic space involves generalized white noise functionals, each of which can be expressed as a sum of homogeneous chaos(es) so as to be fitting for our analysis, where the time or the space-time evolution appears explicitly (such an analysis is the so-called *causal calculus*). Similar situations are seen in the analysis of generalized Poisson noise functionals and even for functionals of compound Poisson noise. All come from the fact that a noise, an idealized random variable, is taken to be the variable of functionals.
(2) The infinite dimensional rotation group plays an important role, and hence our analysis has a side of an infinite dimensional harmonic analysis. Essentially infinite dimensional property can be illustrated by this rotation group.
(3) The innovation approach in terms of white noise is the key theory to the study of complex system, the most typical case is concerned with random complex systems, in particular random fields which are our main research objects.

Chapter 2

White Noise

In this chapter, some necessary background of the theory of white noise analysis will quickly be summarised to have a short excursion of the basic theory for our purpose. For further background the reader is recommended to see the literatures [28] and [42].

2.1 Preliminaries

It is important to note that our idea of white noise analysis starts out with a step of the reduction. Namely, let a random complex system which is to be analyzed be given, then we try to find a system of *idealized elemental random variables* (i.e.r.v.), as the step of reduction, so that the given system is expressed as a functional of the i.e.r.v.'s. In most favourable cases, they form a generalized stochastic process with independent values at every point.

The familiar way of the analysis of functionals of the independent elemental random variables is to introduce a Hilbert space spanned by those functionals with finite variance, and appeal to the standard way of the analysis on Hilbert space. This will be discussed below systematically. Here is an important remark that there are many interesting functionals that can not be dealt with by the Hilbert space technique, i.e. L^2-theoretical methods. Of course, the method to discuss those functionals has a lot of varieties, however we shall show some standard cases, since it is difficult to cover all the cases.

In many important cases the i.e.r.v. is taken to be a white noise; more precisely a Gaussian white noise. To fix the idea, the parameter space of the white noise is now taken to be R. The analysis goes on the Hilbert space (L^2) which is defined by the steps (1) and (2) below.

(1) Heuristically speaking, we take time derivative $\dot{B}(t) = \frac{d}{dt}B(t)$ of a Brownian motion $B(t)$ which has stationary independent increments, so that we are given a system $\{\dot{B}(t)\}$ of idealized elemental random variables. The $\{\dot{B}(t)\}$ is a generalized stochastic process which is stationary. To speak of its probability distribution we compute the characteristic functional $C(\xi) = E[e^{i\langle \dot{B},\xi\rangle}], \xi$ being a test function in E involving smooth functions on R.
(2) It is easy to see that $C(\xi) = \exp\left[-\frac{1}{2}\|\xi\|^2\right]$, where $\| \ \|$ is $L^2(R^1)$-norm. Now take E to be a nuclear space dense in $L^2(R^1)$. The $C(\xi)$ is continuous, positive definite and $C(0) = 1$. Hence, we can appeal to the Bochner–Minlos theorem (see Appendix 1).

Let $(E^*, \mathcal{B}, \mu)$ be given, where E^* is a space of generalized functions on R, more precisely E^* is the dual space of $E \subset L^2(R^1)$, $\mathcal{B}$ is a σ-field generated by cylinder subsets of E^* and where μ is the Gaussian measure on $(E^*, \mathcal{B})$ such that its characteristic functional $C(\xi)$, $\xi \in E$, is given by

$$C(\xi) = \int_{E^*} \exp[i\langle x, \xi\rangle] d\mu(x) \\ = \exp\left[-\frac{1}{2}\|\xi\|^2\right], \quad \xi \in E. \tag{2.1.1}$$

Remark 2.1 *The system $\{\dot{B}(t)\}$ is a white noise. We also call the measure space $(E^*, \mathcal{B}, \mu)$ a white noise. It is, in fact, a realization of white noise. Originally, a white noise is understood as a system of independent infinitesimal random variables which are time independent and Gaussian in distribution like $\{\dot{B}(t)\}$ given above. We may call even $x(\cdot)$ in (E^*, μ) a white noise.*

Then, the complex Hilbert space

$$(L^2) = L^2(E^*, \mu) \\ = \{\varphi(x);\ \text{complex valued},\ \mu\text{-square integrable}\} \tag{2.1.2}$$

can be built in the usual manner, and it is taken to be the basic space of *white noise analysis*.

Starting from (L^2) we can construct a Gel'fand triple

$$(S) \subset (L^2) \subset (S)^*, \tag{2.1.3}$$

where (S) and $(S)^*$ are the space of test functionals and that of generalized (white noise) functionals, respectively. Intuitively speaking, the above triple is an infinite dimensional analogue of the triple

$$S \subset L^2(R^1) \subset S' \tag{2.1.4}$$

in the definition of the Schwartz space of distributions.

An actual construction of the space $(S)^*$ of generalized white noise functionals is shown successively in what follows.

The canonical bilinear form $\langle x, \xi \rangle$ connecting E^* and E, $x \in E^*, \xi \in E$, is $\mathcal{B}$-measurable function of x for a fixed $\xi \in E$, that is a random variable on (E^*, μ). Indeed, it is a Gaussian random variable with mean zero and variance $\|\xi\|^2$, and hence it is a member of (L^2). Suppose a sequence $\{\xi_n, n \in N\}$, with $\xi_n \in E$ for every n, and converges to f in $L^2(R^d)$. Then, $\{\langle x, \xi_n \rangle\}$ is a Cauchy sequence and so $\langle x, \xi_n \rangle$ converges in (L^2). The limit is denoted by $\langle x, f \rangle$ which is a random variable determined by f, almost everywhere. Such a functional $\langle x, f \rangle$ is called a stochastic bilinear form.

Further it is possible to define the bilinear form taking f to be a generalized function, e.g. a delta function to have $x(t)$, and even the polynomial in $x(t)$'s can be defined. For this purpose, we explain the following.

The multi-dimensional parameter (say R^d-parameter) case can be discussed in exactly the same manner to the case R^1, where $L^2(R^1)$ is replaced by $L^2(R^d)$. We often use a simple notation like $x(u), u \in R^d$, to express a white noise.

Wick product

Let $x = x(u)$ be a white noise. The n-th order Wick product $: x^{\otimes n} :$, more precisely $: x^{\otimes n}(u_1, u_2, \ldots, u_n) =: x(u_1)x(u_2)\cdots x(u_n) :$ of white noise is defined by the following formal recursive formula.

$$: x^{\otimes 0} : = 1 \tag{2.1.5}$$

$$: x^{\otimes 1} : = x \tag{2.1.6}$$

$$: x^{\otimes n} : = : x^{\otimes n-1} : \hat{\otimes} x - (n-1) : x^{\otimes n-2} : \hat{\otimes} Tr, \tag{2.1.7}$$

where $Tr \in \widehat{S'(R^2)}$ is the generalized function given by

$$Tr := \int \delta_t^{\otimes 2} dt,$$

i.e., if $f \in S(R^2)$, then $\langle Tr, f \rangle = \int f(t,t)dt$. The symbol $\hat{}$ means symmetrization.

It follows that generalized function : $x_{t_j}^{\otimes n}$:=: $x(t_1)x(t_2)\cdots x(t_n)$: $(t_1, t_2, \ldots, t_n) \in R^n$, belongs to the space $\widehat{S'(R^n)}$. With this understanding the above recursive formula can be written as

$$:1: = 1 \tag{2.1.8}$$

$$:x(u_1): = x(u_1) \tag{2.1.9}$$

$$:x(u_1)x(u_2): = x(u_1)x(u_2)\delta(u_1 - u_2)$$

$$:x(u_1)\cdots x(u_n): = :x(u_1)\cdots x(u_{n-1}):x(u_n) \tag{2.1.10}$$

$$- \sum_{i=1}^{n-1} \delta(u_n - u_i) : x(u_1)\cdots x(u_{i-1})x(u_{i+1})\cdots x(u_{n-1}):,$$

respectively.

We now come back to the basic Hilbert space (L^2), for which we can form the **Fock space**.

Theorem 2.1 *We have a direct decomposition:*

$$(L^2) = \bigoplus H_n,$$

where the subspace H_n is spanned by the Hermite polynomials in the white noise $x(u)$ of degree n.

The Hermite polynomials are nothing but the Wick product denoted by $: x(u_1)x(u_2)\cdots x(u_n) :$, and more explicitly the subspace H_n is the collection of white noise functionals of the form

$$\int \cdots \int_{R^n} F(u_1, u_2, \ldots, u_n) : x(u_1)x(u_2)\cdots x(u_n) : du^n, \tag{2.1.11}$$

where F is a symmetric $L^2(R^n)$-function.

The above functional of x is identified with the *homogeneous chaos* of degree n after N. Wiener, or we may call the *multiple Wiener integrals* of degree n introduced by K. Itô.

Let $\widehat{L^2(R^n)}^{d\otimes}$ be the subspace of $L^2(R^{n\times d})$ involving square integrable functions $F(u_1, u_2, \ldots, u_n), u_j \in R^d, j = 1, 2, \ldots, n$, which are symmetric in n variables u_j's. Then we have the following theorem for R^d-parameter case.

The space (L^2) and the Fock space are easily generalized to the case $d > 1$.

Theorem 2.2 [18] (*Integral representation*)
There is a bijection between H_n *and* $\widehat{L^2(R^n)}^{d\otimes}$, *such that*

$$\varphi \leftrightarrow F \in \widehat{L^2(R^n)}^{d\otimes}, \quad \varphi \in H_n$$

and

$$\|\varphi\|_{(L^2)} = n!\|F\|^2_{L^2(R^{n\times d})}.$$

It is important to note that any (L^2)-functionals can be represented by a sequence F_n of symmetric $L^2(R^{n\times d})$-functions defined on *finite* dimensional Euclidean space. The system of basic random variables is $\{x(u), u \in R^d\}$. Given such a system, then the most elementary function φ is a polynomial in the $x(u)$, but it is not defined in the ordinary calculus since $x(u)$ is a generalized function of u. However, we may define a linear functional of the tensor product of the $x(u)$'s with some renormalization. It is nothing but the above integral (2.1.11).

It would be reasonable to take $x(t)$ without smearing to describe the evolutional random phenomena having the time variable explicitly. The white noise functionals in terms of $x(u)$'s are necessarily not ordinary (L^2)-functionals but generalized functionals.

We are now ready to introduce the spaces of generalized white noise functionals. Actually there are two typical types of a space of such generalized functionals. To fix the idea we consider the case $d = 1$.

I. Space $(S)^*$

Let

$$A = -\frac{d^2}{du^2} + u^2 + 1. \tag{2.1.12}$$

For any $\varphi \in (L^2)$, by using Theorem 2.1 and Theorem 2.2, it can be uniquely expressed as

$$\varphi = \sum_{n=0}^{\infty} \varphi_n, \quad \varphi_n \in H_n,$$

and can have a representation

$$\varphi_n \leftrightarrow F_n \in \widehat{L^2(R^n)}.$$

Thus, we associate a sequence F_n to any $\varphi \in (L^2)$. For any φ, represented above, satisfying the condition

$$\sum_{n=0}^{\infty} n! |A^{\otimes n} F_n|_0^2 < \infty,$$

we define $\Gamma(A)\varphi \in (L^2)$ by

$$\Gamma(A)\varphi \leftrightarrow \sum_{n=0}^{\infty} (A^{\otimes n} F_n).$$

This operator $\Gamma(A)$ is called the *second quantization operator* of A. We can see that

1. $\Gamma(A)$ has a set of eigenfunctions which forms an orthonormal basis for (L^2).
2. $\Gamma(A)^{-1}$ is a bounded operator acting on (L^2) with $\|\Gamma(A)^{-1}\| = 1$.
3. For any $p > 1$, the operator $\Gamma(A)^{-p}$ is of Hilbert–Schmidt type.

For each $p \geq 0$, define

$$\|\varphi\|_p = \|\Gamma(A)^p \varphi\|_0,$$

where $\|\cdot\|_0$ is the (L^2)-norm. Let

$$(S)_p \equiv \{\varphi \in (L^2); \|\varphi\|_p < \infty\}.$$

Then $(S)_p$ is a Hilbert space with $\|\cdot\|_p$. Define

$$(S) \equiv \text{projective limit of } \{(S)_p; p \geq 0\}.$$

Then (S) is a nuclear space by the above property 3 of $\Gamma(A)$. We call (S) a space of test functions. The dual space $(S)^*$ of (S) is called the space of generalized white noise functionals. Thus we have a Gel'fand triple:

$$(S) \subset (L^2) \subset (S)^*$$

as is announced in (2.1.3).

II. Space $(L^2)^-$

First extend the isomorphism between H_n and $\widehat{L^2(R^n)}$ established in Theorem 2.2 ($d = 1$) to have

$$H^{-\frac{n+1}{2}}\widehat{(R^n)} \cong H_n^{-n},$$

where $\widehat{H^m(R^n)}$ is the subspace of the Sobolev space $H^m(R^n)$ of order m, consisting of symmetric functions on R^n. The direct sum

$$(L^2)^- = \bigoplus_{n=0}^{\infty} c_n H_n^{-n}$$

is a space of generalized white noise functionals, where $\{c_n\}$ is chosen depending on the purpose, in such a way that $c_n > 0$ for every n and $c_n \to 0$.

Note that $(L^2)^-$ has a Hilbert space structure.

Example 2.1 Generalized white noise functionals

1. $x(t) = \dot{B}(t)$ is a generalized function of x. Note that in the elementary calculus it is not permitted to evaluate a generalized function x rigorously at t but we can do.
2. Renormalized polynomials (Wick product) are of the form

$$: \prod_j x(t_j) : .$$

3. The Gauss kernel

$$\varphi_c(x) = \mathcal{N} e^{c \int x(t)^2 dt}, \quad c \neq \frac{1}{2},$$

where $\mathcal{N}$ is a renormalizing constant.

S-transform

The representation (2.1.11) in terms of symmetric $L^2(R^n)$-functions comes from the S-transform given below. The *S-transform* of a generalized functional $\varphi(x) \in (S)^*$ is defined by

$$(S\varphi)(\xi) = \int \varphi(x + \xi) d\mu(x), \quad \varphi \in (S)^*, \tag{2.1.13}$$

which may be viewed as an infinite dimensional Laplace transform, since we can establish

$$(S\varphi)(\xi) = \exp\left[-\frac{1}{2}\|\xi\|^2\right] \int \exp[\langle x, \xi \rangle]\varphi(x) d\mu(x). \tag{2.1.14}$$

The $(S\varphi)(\xi)$ is often called U-functional.

Example 2.2 Continuation of Example 2.1.

The S-transform of generalized functionals in Example 2.1 are

1. $(Sx)(\xi) = \xi(t)$,
2. $(S : \prod_j x(t_j) :)(\xi) = \prod_j \xi(t_j)$,
3. $(S\varphi_c)(\xi) = e^{\frac{c}{1-2c}\int \xi(t)^2 dt}, c \neq \frac{1}{2}$

respectively.

These examples, together with others, illustrate that the S-transform gives visualized expression and that renormalization is not ad hoc, but quite natural.

Remark 2.2 *Since φ is a generalized functional, it should be noted that φ can be defined for the variable $x + \xi$ which is still in the support of μ and that the integral* (2.1.13) *is understood to be the value of $\varphi(\cdot + \xi)$ evaluated at 1.*

There is a characterization theorem for generalized white noise functionals. The theorem is powerful when we discuss analysis on U-functional in place of generalized functional.

Theorem 2.3 (*Potthoff–Streit*)
Let $\Phi \in (\mathcal{E})^*_\beta$. Then its S-transform $F = S\Phi$ satisfies the conditions

(a) For any ξ and η in $\mathcal{E}_p$, the function $F(z\xi + \eta)$ is an entire function of $s \in \mathbf{C}$.
(b) There exists non-negative constants K, α, and p such that

$$|F(\xi)| \leq K \exp\left[\alpha|\xi|_p^{2/(1-\beta)}\right], \quad for\ every\ \xi \in \mathcal{E}_c.$$

Conversely, suppose a function F defined on $\mathcal{E}_p$ satisfies the above two conditions. Then there exists a unique $\Phi \in (\mathcal{E})^*_\beta$ such that $F = S\Phi$ for any q satisfying the condition that $e^2(\frac{2\alpha}{1-\beta})^{1-\beta}\|A^{-(q-p)}\|^2_{HS} < 1$, the following inequality holds:

$$\|\Phi\|_{-q-\beta} \leq K\left(1 - e^2\left(\frac{2\alpha}{1-\beta}\right)^{1-\beta}\|A^{-(q-p)}\|^2_{HS}\right)^{-1/2}.$$

(*See* [42], *Chapter 8.*)

There is no problem to define the formula (2.1.14) for φ in $(S)^*$, since the exponential function is in (S). Thus, by using the S-transform, a *differential* *operator* ∂_t, acting on (S), can be defined as follows

$$\partial_t = S^{-1}\left(\frac{\delta}{\delta\xi(t)}(S\varphi)(\xi)\right), \tag{2.1.15}$$

where $\frac{\delta}{\delta\xi(t)}$ denotes the Fréchet derivative.

Note that a strong topology is introduced to the space E so as the Fréchet derivatives are defined rigorously.

Remark 2.3 *For the one dimensional parameter case, a concretized expression of white noise is $\dot{B}(t)$ which is the time derivative of a Brownian motion $B(t)$, and there μ-almost all $x \in E^*$ are viewed as sample paths of $\dot{B}(t)$. We may also regard the operator ∂_t as the partial differential operator $\frac{\partial}{\partial\dot{B}(t)}$ in the variable $\dot{B}(t)$, which has been defined rigorously by* (2.1.15). *Here we note that the S-transform of $\dot{B}(t)$ is $\xi(t)$.*

The differential operator ∂_t acts as an *annihilation operator.* While the *creation operator* ∂_t^* is defined on $(S)^*$ as the adjoint operator of ∂_t in such a way that

$$\langle \partial_t\varphi, \psi\rangle = \langle \varphi, \partial_t^*\psi\rangle, \quad \varphi \in (S), \ \psi \in (S)^*, \tag{2.1.16}$$

which plays an important role in an infinite dimensional stochastic analysis, e.g. a stochastic integral, indeed Hitsuda–Skorokhod integral and others.

The sum

$$m_t = \partial_t + \partial_t^*$$

is the multiplication by $x(t)$. At the same time it stands for a *quantum white noise.* Through this fact, we can see good connection with the quantum probability theory.

Note. The role of ∂_t is quite different from that of $\frac{d}{dt}$ (or $\frac{\delta}{\delta C}$). Sometimes their roles are mutually complementary, and other times they are used together in our calculus. The ∂_t and hence ∂_t^* can appear only in the stochastic calculus. This fact illustrates the complexity of the analysis of random function.

In each case we have a *graded algebra* under the multiplication defined by Wick product. Creation and annihilation operators act to increase and to decrease the grade respectively. Such an understanding is helpful for the application of white noise theory to other fields where non-commutative operators are used (e.g. [43]).

$\mathcal{T}$-transform

The $\mathcal{T}$-transform was introduced earlier than the S-transform, which is now used more frequently by many reasons. The $\mathcal{T}$-transform is defined by

$$(\mathcal{T}\varphi)(\xi) = \int_{E^*} \exp[i\langle x, \xi\rangle]\varphi(x)d\mu(x), \quad \varphi \in (L^2),$$

which can be thought of an infinite dimensional analogue of the Fourier transform. Indeed, it plays some roles similar to the finite dimensional Fourier transform. If one wishes to define a transformation that enjoys more similar properties, then he can see the Fourier–Wiener transform (see [18]). Here we only expect a transformation that gives some visualized representation of the white noise functionals. The $\mathcal{T}$-transform, like S-transform, complies with our hope.

If φ is in the subspace H_n of homogeneous chaos of degree n, then we can prove

$$(\mathcal{T}\varphi)(\xi) = i^n C(\xi)U(\xi), \quad \xi \in E,$$

where $U(\xi)$ is the S-transform of φ, and is homogeneous in ξ of degree n. The U is called U-functional associated to φ.

The following assertions can easily be proved.

Proposition 2.1

(1) *If φ is an integrable white noise functional, then its $\mathcal{T}$-transform is defined and is continuous in $\xi \in E$.*

(2) *The domain of the $\mathcal{T}$-transform extends to the space (S^*) of generalized white noise functionals.*

Because of the property (1), we often use the $\mathcal{T}$-transform, as we shall see in Section 3.9 to define the space $(\mathbf{P})$.

The assertion (2) follows from the fact that $\exp[i\langle x, \xi\rangle]$ plays a role of test functional.

2.2 Multi-dimensional parameter white noise

Let $x(u), u \in R^d$, denote an R^d-parameter white noise. Our main idea is to restrict the parameter u to C which is a smooth, convex and closed manifold in R^d. In particular, C is often taken to be a smooth ovaloid. This choice is important when the variational calculus is discussed. Before we come to this topic, some observation on related topics is necessary.

It is noted that the restrictions of the parameter should always be consistent. Observation on the parameter restriction of the Lévy Brownian motion as well as the relationship with the white noise would be helpful for a study of parameter restriction of white noise itself. If the parameter of the Lévy Brownian motion with R^d parameter is restricted to be in a lower dimensional hyperplane, say $R^{d'}$ with $d' < d$, passing through the origin, then we are given $R^{d'}$ parameter Brownian motion. This is one of the reason why the Lévy Brownian motion is a natural generalization of the ordinary Brownian motion. However, it is not so simple to make a generalization of the obvious relationship between Brownian motion $B(t)$ and white noise $\dot{B}(t)$. A multi-dimensional Lévy's Brownian motion describes more complex dependence among $B(t), t \in R^d$, so that the relationship between the Brownian motion and the white noise is much complicated (see H.P. McKean [56]).

So far the parameter space of white noise has been taken to be the whole R^d. It is easy to define a white noise with parameter restricted to a domain, contained in R^d. However, we have to be careful when we let the parameter be restricted to a lower dimensional manifold M, contained in R^d. When we discuss variational calculus, we usually meet differentials where the parameter of the white noise is restricted to a surface or a curve. There we tacitly assume the possibility of such a restriction, but not arbitrarily. It is necessary to clarify this fact.

Assume that a white noise $(E^*, \mu), E^* \subset L^2(R^d)$, with a parameter space R^d is given in advance.

Proposition 2.2 *Let M be an ovaloid in R^d of C^∞-class. Then, a white noise measure with parameter set M is defined, and it is in agreement with the one obtained from the original white noise measure by the restriction of the parameter to M.*

Proof. The characteristic functional $C(\xi)$, given by (2.1.1), is written in the form

$$C(\xi) = \exp\left[-\frac{1}{2}\int_{R^d} \xi(u)^2 du\right]. \tag{2.2.1}$$

The integral on R^d in the above formula can be restricted to that on a smooth manifold M to have

$$C_M(\xi) = \exp\left[-\frac{1}{2}\int_{M} \xi(u)^2 du\right], \tag{2.2.2}$$

where ξ restricted to M is in $E(M)$ (a nuclear space of C^∞-functions on M) and where du is the volume element on the manifold M. Thus a Gel'fand triple

$$E(M) \subset L^2(M) \subset E(M)^* \tag{2.2.3}$$

is obtained and we are given a Gaussian measure μ_M on $E^*(M)$ which is uniquely determined by $C_M(\xi)$. Note that the construction of the above Gel'fand triple heavily depends on the differential structure of the manifold M. Indeed, it is possible to introduce a nuclear space $E(M)$ with a suitable choice of M so that the Bochner–Minlos theorem holds in order to guarantee the existence of a probability measure μ_M on $E(M)^*$.

It is noted that such a restriction of parameter is consistent with the choice of a manifold M like orthogonal projection in Hilbert space. Namely, if M' is another manifold satisfying the regular conditions and if $M' \subset M$, then the probability measure μ' is the probability constructed from μ_M and so is from μ itself.

Proposition 2.3 *The measure μ_M is viewed as an induced distribution of the original white noise measure μ.*

It can be easily proved by noting what is mentioned just before this proposition and since white noise has independent values at every point, we omit the proof.

This fact will tacitly be used later, where a manifold M is specified to be a contour C or a surface.

We now note that the actions by the operators ∂_t and ∂_t^* can be restricted to the spaces $(S(M))$ and $(S(M))^*$, respectively, for any convex C^∞-manifold M as was specified before so that a Gel'fand triple of functions on M can be defined.

Note. It is interesting to study a random field $X(a), a \in R^d$, when a is increasing in a radial direction. We can actually form such a field as an increasing stable noise random measure. This field can be used as a random multi-dimensional time.

Finally, we come to the restriction of the parameter of white noise to a lower dimensional hyperplane or a manifold. In such a case, we need to use the generalized white noise functionals supported by the hyperplane or the manifold in question.

Restriction of the parameter to lower dimensional domain

(1) Gaussian case

Given an R^d parameter white noise (E^*, μ). For $f \in R^d$ the stochastic bilinear form $\langle x, f\rangle, x \in E^*$, is defined. It is subject to a Gaussian distribution $N(0, \|f\|^2)$. Take f to be an indicator function such that

$$f_t(u) = \chi_{I(t)}(u), \quad t = (t_1, t_2, \dots, t_d), \quad I(t) = \Pi_j[0, t_j].$$

A Brownian sheet is an R^d_+-parameter Gaussian system $\{W(t), t = (t_1, t_2, \dots, t_d) \in R^d_+\}$ such that

(i) $E(W(t)) = 0$,

(ii) $E(W(t)W(s)) = \prod_1^d \min(t_j, s_j)$, where R^d_+ denotes the non-negative quadrant in R^d.

Brownian sheet is one of the generalizations of a standard Brownian motion depending on $t \in R^1$.

Various properties including the Markov property have been discussed extensively. (See, e.g. R. C. Dalang and J. B. Walsh [6].)

We have done only in the place where (Gaussian) white noise is actually constructed. Similarly, the Poisson sheet can be defined.

Set $t_d \equiv 1$. Then, we are given an R^{d-1}-dimensional parameter Brownian sheet. It is now ready to have an R^{d-1}-dimensional parameter white noise by applying partial derivatives in t_j's $(j \neq d)$.

If one is concerned only with the probability distribution, then he can take a characteristic functional as before and can let the variable function $\xi(u), u \in R^d$, be restricted to a hyperplane to have the ξ restricted: $\xi(u)|_{u \in R^{d-1}}$, which is to be the new variable of the characteristic functional in question. Note that this is quite different from the restriction of vector valued random variables. This remark is meaningful when variation of a random field is discussed.

Similarly, much lower dimensional parameter Brownian sheet and hence, lower dimensional parameter white noise can be derived.

The idea of restricting the parameter to a hyper surface C is the same for the case of restricting to a hyperplane provided that the C is assumed to be a smooth ovaloid. Taking vectors with unit length along the normal direction to have $C + \delta C$ and integrate the white noise over the region enclosed by C and $C + \delta C$. Then, a white noise parameterized by a point s on C is obtained. Note that such a system of white noise associated

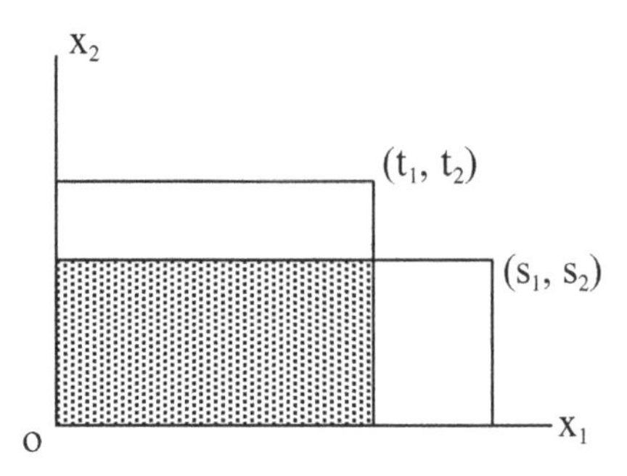

Fig. 2. Brownian sheet.

with C is consistent, since every one of them in the system comes from the higher-dimensional parameter white noise given in advance.

Proposition 2.4 *The restriction of the parameter of white noise can be done with the help of Brownian sheet viewed as a random field.*

Another method of restricting the parameter will be discussed later in Section 7.2.

(2) Poisson case

For the Poisson noise the same trick can be applied so far as the Hilbert space method is concerned. It will be discussed in Section 3.1.

The path space theoretical approach to a Poisson noise will come later, where somewhat different type of the probabilistic properties will be observed.

2.3 Infinite dimensional rotation group $O(E)$

Leaving the theory of white noise functionals for a moment we now introduce the infinite dimensional rotation group. The effective use of the group for the calculus is one of the big advantages of white noise analysis. Heuristically, the infinite dimensional rotation group was originally introduced by H. Yoshizawa in 1969 (published 1970). One might be surprised that the term "rotation group" appears here, however, as will be seen in what follows, this group plays a fundamental role in white noise analysis, as it were, an infinite dimensional harmonic analysis. In fact, an invariance of the white noise measure is described by the rotation group, the irreducible representations of the group is used to explain important properties of Brownian motion, the action of the conformal subgroup enables us to have the innovation from the variation of a random field, and so forth.

Now the definition is given. Let E denote a basic nuclear space which is a dense subspace of $L^2(R^d)$.

Definition 2.1 A linear transformation g is called a *rotation* of E if it satisfies

(i) a homeomorphism of E,
(ii) an orthogonal transformation acting on E, i.e. $\|g\xi\| = \|\xi\|$ for every $\xi \in E, \|\cdot\|$ being the $L^2(R^d)$-norm.

Obviously, the collection of all rotations of E forms a group, which is denoted by $O(E)$. The compact-open topology is naturally introduced, since $O(E)$ is a transformation group acting on a topological space E. Thus, we are given a topological group.

Definition 2.2 The topological group $O(E)$ is called the *rotation group* of E.

Remark 2.4 *The nuclear space E can be the Schwartz space S, in most cases. However, when the conformal group is imbedded in $O(E)$, it is convenient to take E to be $D_0(R)$ which is diffeomorphic to $C^\infty(S^1)$. This fact will be discussed in Section 10.1.*

There is much freedom to choose the basic space E in order to introduce a rotation group. In case the basic nuclear space is not explicitly stated, the rotation group is often denoted by O_∞.

It is noted that the group $O(E)$ for any choice of E is neither compact nor locally compact. We therefore take suitable subgroups depending on the purpose. Those subgroups play significant role in white noise analysis. Therefore, we may say that the white noise analysis has an aspect of an infinite dimensional *harmonic analysis* arising from the infinite dimensional rotation group.

Associated with a rotation g is the adjoint operator g^* which is uniquely determined in the usual manner:

$$\langle x, g\xi \rangle = \langle g^*x, \xi \rangle, \quad x \in E^*, \ \xi \in E,$$

where $\langle \cdot, \cdot \rangle$ is the canonical bilinear form that connects E and E^*. The collection $O^*(E^*) = \{g^*; \ g \in O(E)\}$ forms a group as is easily seen. In addition, we can prove the following proposition.

Proposition 2.5 *The group $O^*(E^*)$ is isomorphic to $O(E)$ by the corresponding*

$$g \leftrightarrow (g^*)^{-1}, \quad g \in O(E),$$

which is bijective.

Because of this assertion, the group $O^*(E^*)$ is also called the infinite dimensional rotation group. To avoid confusion, we may say rotation group of E^*.

Now the key theorem can be presented. Remind that we fix the measure space $(E^*, \mathcal{B}, \mu)$ of white noise.

Theorem 2.4 *For any g^* in $O^*(E^*)$*

(i) *g^* is $\mathcal{B}$-measurable, and*
(ii) *white noise measure μ is g^*-invariant:*

$$dg^*\mu(x) = d\mu(x),$$

where $dg^*\mu(x)$ is defined to be $d\mu(g^*x)$.

Proof. The assertion (i) follows from the fact that g^* is bijective from the algebra of cylinder sets onto itself, so g^* is $\mathcal{B}$-measurable .

As for the proof of (ii), we observe the characteristic functional of $g^*\mu$:

$$\begin{aligned}\int_{E^*} \exp[i\langle x, \xi\rangle] d\mu(g^*x) &= \int_{E^*} \exp[i\langle g^{*-1}y, \xi\rangle] d\mu(y) \\ &= \int_{E^*} \exp[i\langle y, g^{-1}\xi\rangle] d\mu(y) \\ &= \exp\left[-\frac{1}{2}\|g^{-1}\xi\|^2\right] \\ &= \exp\left[-\frac{1}{2}\|\xi\|^2\right], \end{aligned} \tag{2.3.1}$$

which is agreement with that of μ.

Note. It can further be proved that the measure μ is $O^*(E^*)$-*ergodic.* This means that if A is a $\mathcal{B}$-measurable set invariant under $O^*(E^*)$, i.e. $\mu(g^*A) = \mu(A)$ for every $g^* \in O^*(E^*)$, then $\mu(A) = 0$ or 1.

2.4 Subgroups of $O(E)$

As will be seen in Fig. 3, probabilistically interesting subgroups of $O(E)$ will be divided into two parts (in the Fig. 3, upper half part and lower half part). The first part, denoted I, the definition of subgroups comes from the choice of a complete orthonormal system $\{\xi_n\}$ of $L^2(R^n)$, where each ξ_n is a member of E. The part I contains the subgroups (1), (2) and (3) listed below. While the second part, denoted by II, involves members which is defined by the diffeomorphisms. Hence the transformations acting on E defined by the subgroups in II are coordinate-free. The subgroup (4) below belongs to the class II.

(1) Finite dimensional rotations

Take an increasing sequence of subspaces E_n, $\dim E_n = n$.
Members of $O(E)$ such that

$$g|_{E_n} \in SO(n)$$

and

$$g|_{E_n^\perp} = I, \quad I : \text{identity}$$

form a subgroup of G_n of $O(E)$.

The inductive limit G_∞ of G_n is again a subgroup of $O(E)$. This group characterizes the infinite dimensional Laplace–Beltrami operator.

Here is a method of the characterization, which is rather well known. First the algebra generated by the generators of two-dimensional rotations. Once the coordinate system $\{\xi_n\}$ of E is fixed, the generators can be expressed in the form

$$\gamma_{j,k} = \xi_j \frac{\partial}{\partial \xi_k} - \xi_k \frac{\partial}{\partial \xi_j}, \quad 1 \le j \ne k < \infty.$$

Form such a quadratic form Q of the $\gamma_{j,k}$'s as

(1) Q commutes with all $\gamma_{j,k}$'s,
(2) Q annihilates constants,
(3) Q is positive: namely,

$$(Q\xi, \xi) \ge 0.$$

Then, Q is expressed in the form

$$Q = c\Delta_\infty, \quad c \le 0,$$

where

$$\Delta_\infty = \sum_1^\infty \left(\frac{\partial^2}{\partial \xi_n^2} - \xi_n \frac{\partial}{\partial \xi_n} \right). \tag{2.4.1}$$

Definition 2.3 The operator Δ_∞ is called the infinite dimensional Laplace–Beltrami operator.

Remark 2.5 *In quantum dynamics $\mathcal{N} = -\Delta_\infty$ is called the* **number operator**, *since H_n in the Fock space is viewed as the collection of states where n Bose particles are there and*

$$\mathcal{N}\varphi = n\varphi, \quad for\ every\ \varphi \in H_n.$$

It is known that (see, e.g. [28], [42]) the domain of Δ_∞ is dense in (L^2), and in particular, the subspace H_n is the eigenspace:

$$\Delta_\infty \varphi = -n\varphi, \quad \varphi \in H_n.$$

Where the chaos expansion is available, the operator Δ_∞ can be used efficiently. Namely, a member of H_n is characterized as an eigen function of the operator Δ_∞.

(2) The Lévy group

Let π be an automorphism of the set Z^+ of positive integers. The *density* of π is defined by

$$d(\pi) = \limsup_{N\to\infty} \frac{1}{N} \#\{n \le N; \pi(n) > N\},$$

where $\#$ denotes the number of elements in $\{\ \}$.

Let g_π be defined in such a way that

$$\xi = \sum a_n \xi_n \to g_\pi \xi = \sum a_n \xi_{\pi(n)}.$$

Set

$$\mathcal{G} = \{g_\pi;\ d(\pi) = 0,\ \ g_\pi \in O(E)\}.$$

Then, $\mathcal{G}$ forms a subgroup of $O(E)$. The group $\mathcal{G}$ has a close connection with the Lévy Laplacian Δ_L. This fact will be prescribed separately.

(3) The group $\mathcal{H}$ of changing sign of ξ

The binary expansion of $t \in [0, 1]$ is given by

$$t = \sum_0^\infty \eta(n) 2^{-n}, \quad \eta(n) = 0 \text{ or } 1.$$

Set $\epsilon(n) = 2\eta(n) - 1$. Then, $\epsilon(n)$ is -1 or 1. We have

$$t \leftrightarrow \{\epsilon(n)\} \cong \{-1, 1\}^\infty.$$

Let $\{\xi_n\}$be a complete orthonormal system. Define

$$g_t : \xi = \sum a_n \xi_n \leftrightarrow g_t \xi = \sum a_n \epsilon(n) \xi(n).$$

Then, obviously

$$\|g_t \xi\|^2 = \sum a_n^2 = \|\xi\|^2$$

holds.

(i) For any t, the g_t is continous and linear on E.
This is proved by the equality

$$\|g_t \xi - g_t \eta\|^2 = \sum (a_n - b_n)^2 \epsilon(n)^2 = \|\xi - \eta\|^2.$$

(ii) The g_t is continous in t and $g_0 = e$ (identity).
If $|t-s|$ is small, the equality $\epsilon(n) = \epsilon'(n),\ n \geq N$, holds for sufficiently large N. Hence, we have

$$\|g_t \xi - g_s \xi\|^2 \leq 4 \sum_N^{\infty} a_n^2.$$

In particular

$$g_t \to e : \text{ identity as } t \to 0.$$

Note. $\{g_t\}$ is not a one-parameter subgroup of $O(E)$.

(iii) Each $g_t, t \neq 0$ is essentially infinite dimensional transformation. In other words, the average power (a.p.)

$$\text{a.p.}(g_t) = \limsup \frac{1}{N} \sum_1^N \|g_t \xi_n - \xi_n\|^2 > 0.$$

(4) Whiskers

A continuous one-parameter group $\{g_t\}$, each member of which comes from diffeomorphisms of the parameter space R^d, is called a *whisker*. More precisely, let $\tilde{g}_t$ be a diffeomorphism of R^d such that

$$g_t : \xi \to (g_t \xi)(u) = \xi(\tilde{g}_t u) |J(\tilde{g}_t u)|^{1/2},$$

where J is the Jacobian and that the following properties hold:

$$\begin{aligned} g_t &\in O(E), \\ g_t g_s &= g_{t+s}, \\ g_t &\to e \quad \text{as} \quad t \to 0. \end{aligned}$$

Particular examples of a whisker can be seen as the conformal group that we shall discuss later in Chapter 9. They serve important roles in the variational calculus for random fields.

The subgroup that is, more general than whisker, involving the members of $O(E)$ that come from diffeomorphism of R^d is particularly important; for example, those members serve to have a variation of a random field $X(C)$ expressed as a functional of white noise when C is deformed by the diffeomorphisms of R^d. There the probability distribution of the innovation is kept invariant. Under those actions the basic white noise is also kept invariant.

The time shift is one of the most important example of a whisker, for which the ergodic property will be discussed.

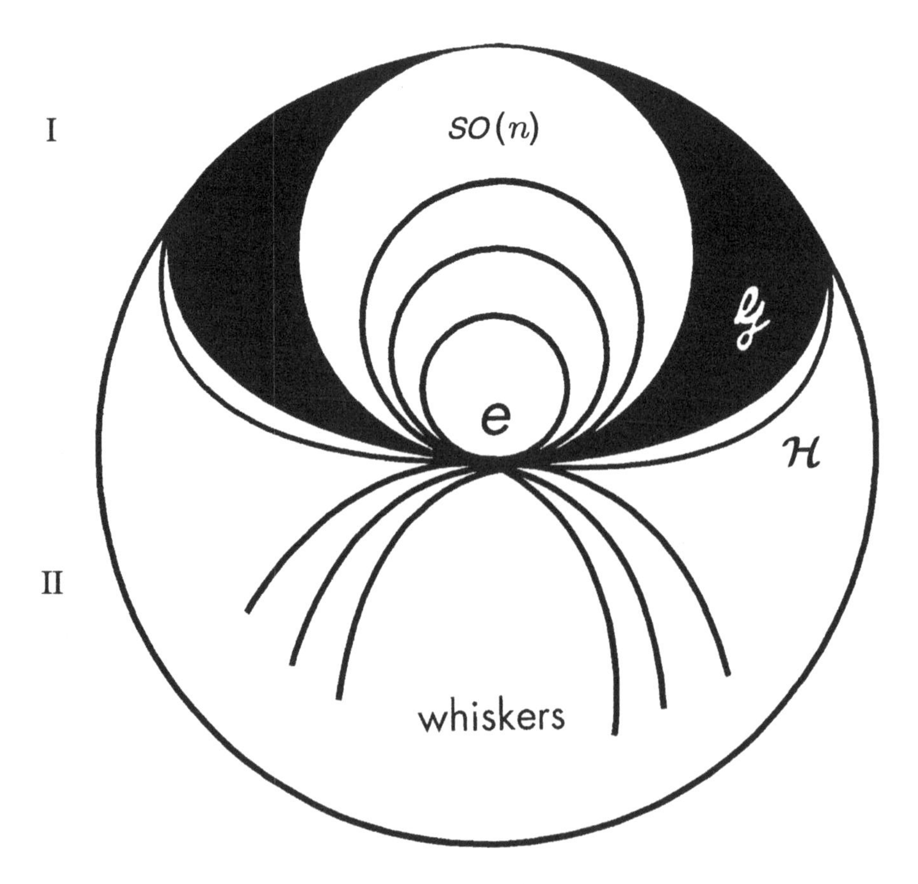

Fig. 3. Infinite dimensional rotation group.

2.5 Laplacians

We have many operators which are analogues of the finite dimensional Laplacian acting on $L^2(R^d)$:

$$\Delta_d = \sum_{i=1}^{d} \frac{\partial^2}{\partial x_i^2}.$$

By using the operators ∂_t and ∂_t^* in Section 2.1, three infinite dimensional Laplacians can be defined. Each plays its own roles, having connections with $O(E)$.

(1) Laplace–Beltrami operator

$$\Delta_\infty = \int \partial_t^* \partial_t dt.$$

An alternative expression has been given in (2.4.1).

(2) Lévy Laplacian

$$\Delta_L = \int \partial_t^2 (dt)^2.$$

The integral is often replaced by $\frac{1}{|T|}\int_T$, T being an interval to have the time. It annihilates members in (L^2), but it effectively acts on $(L^2)^-$ or $(S)^*$.

(3) Volterra Laplacian

$$\Delta_V = \int \partial_t^2 dt.$$

2.6 Invariance of white noise

Invariance of measures associated to white noise under certain transformation will be considered. As for the Gaussian case, there is the infinite dimensional rotation group, discussed in Sections 2.3 and 2.4. We may take another group that gives invariance of white noise measure μ under the transformation of the parameter. Such a group, in addition to whiskers, arises depending on the purpose.

The group $O(E)$ therefore contains not only subgroups which have their own probabilitistic or some other significant meanings, but also individual members that are also useful for optional purpose. For instance, a diffeomorphism of R^d can act on $E \subset L^2(R^d)$ acts in such a way that

$$(g_f \xi)(u) = \xi(f(u))|J_f(u)|^{1/2}$$

defines g_f that is in $O(E)$, where $J_f(u)$ is the Jacobian. Transformations g_f of this type are helpful to define transformations of random fields $X(C)$:

$$g_f X(C) = X(f(C)).$$

Under g_f^* the white noise measure μ is kept invariant, and hence we are given the same field as $X(C)$ in probability distribution.

A transformation of a different type

For the transformation group, there is a group consisting of translations by vectors which are in the support of μ. It is worth to mention the following transformation. Let T_f be a translation of $x \in E^*$ by f.

$$T_f : x \to x + f.$$

Concerning such a translation it is known that

Proposition 2.6 *Let (E^*, μ) be a white noise. The measure μ is quasi-invariant under T_f if and only if f is in $L^2(R^d)$.*

The proof comes essentially from the Kakutani dichotomy (see [18], Section 1.3) for infinite product of probability measures.

Obviously the collection $T = \{T_f, f \in L^2(R^d)\}$ forms a group. Thus, we are led to define the *infinite dimensional group of motions* generated by $O(E)$ and T.

Chapter 3

Poisson Noise

3.1 Poisson noise functionals

A Brownian motion may be said to be one of the elemental additive processes with stationary independent increments and, indeed, most important process. Another elemental additive process is a Poisson process. Let it be denoted by $P(t), t \geq 0$. Remind the probability that $P(t)$ takes the value k (non-negative integer) is $(\lambda^{kt}/k!)e^{-\lambda t}$, where λ is the *intensity*.

The time derivative $\dot{P}(t)$ of a Poisson process is called a *Poisson noise*. A Poisson noise is a generalized stochastic process, with independent values at every t. The time parameter is extended to the entire R, by adding another independent Poisson noise with parameter set $(-\infty, 0]$. We use the same notation $\dot{P}(t)$ for the Poisson noise with R-parameter. Its characteristic functional $C_P(\xi)$, $\xi \in E$, is obtained in the following proposition.

Proposition 3.1 *The characteristic functional $C_P(\xi)$ of $\dot{P}(t), t \in R$, is given by*

$$C_P(\xi) = \exp\left[\lambda \int_R (e^{i\xi(t)} - 1)dt\right]; \quad \xi \in E, \tag{3.1.1}$$

where E is a nuclear space which is dense in $L^2(R)$.

Proof. Let $\{\Delta_j\}$ be a partition of R. Then,

$$\begin{aligned}
E[e^{i\langle \dot{P}, \xi\rangle}] &= \lim_{|\Delta_j| \to 0} E[e^{i\Sigma\xi(t_j)\Delta_j P}], \quad t_j \in \Delta_j \\
&= \lim \prod_j E[e^{i\xi(t_j)\Delta_j P}] \\
&= \lim \exp\left[\lambda \sum (e^{i\xi(t_j)} - 1)|\Delta_j|\right], \quad |\Delta| \text{ being the length of } \Delta, \\
&= \exp\left[\lambda \int_R (e^{i\xi(t)} - 1)dt\right].
\end{aligned}$$

It is often convenient to let it be centered; $\dot{P}(t)-\lambda$, which is a generalized stochastic process with the characteristic functional of the form

$$C_P(\xi) = \exp\left[\lambda \int_R (e^{i\xi(t)} - 1 - i\xi(t))dt\right]. \tag{3.1.2}$$

Now we want to have a generalization of the parameter. Suppose that the characteristic functional $C_P(\xi)$ of a Poisson noise with R^d-parameter is given by

$$C_P(\xi) = \exp\left[\lambda \int_{R^d} (e^{i\xi(t)} - 1)dt^d\right], \tag{3.1.3}$$

as a formal generalization of (3.1.1). Note that we do not have satisfactory definition of R^d-parameter Poisson process (or field). Plausible reasons why we take such a simple generalization to R^d-parameter case will also be given in Section 3.5.

Thus, so far as probability distribution is concerned, the marginal distribution can easily be obtained by the usual method, in particular take the variable t in R^d to be a lower dimensional vector. A probabilistic meaning behind this trick of restriction will also be illustrated in Section 3.5.

Take ξ to be the indicator function of $I(t), t \in R^d$, as in Gaussian case in Section 2.2 and form a stochastic bilinear form $\langle \dot{P}, \xi \rangle$, which is to be a Poisson sheet. A restriction of the parameter to a hyperplane defines a lower dimensional parameter Poisson sheet and then Poisson noise. Also a restriction to a hypersurface is given.

We then come to the analysis of functionals of a Poisson noise. As in the Gaussian case we can state propositions which will be counterparts to the Gaussian case. The technique is quite similar, so that we shall omit the statement.

It is noted that by the observation of a Poisson sheet, we can easily see invariance of Poisson noise under some transformations of the parameter space.

We now introduce a class of functionals of Poisson noise and find that the analysis also has much similarity to the case of Gaussian white noise. Dissimilarity will be explained with much emphasis when we meet it later.

The probability distribution μ_P is introduced on the space E^* which includes $L^2(R^d)$, so that we are given a probability space (E^*, μ_P). The complex Hilbert space $(L^2)_P = L^2(E^*, \mu_P)$ is the basic space, from which our analysis starts.

To fix the idea, we consider the case $d = 1$. By the form of the characteristic functional (3.1.1) it is easily proved that

$$\int_{E^*} \langle x, \xi \rangle^2 d\mu_P(x) = \lambda \int_R \xi(u)^2 du + \lambda^2 \left(\int_R \xi(u) du \right)^2,$$

which is finite. Also, we can prove that

$$\int_{E^*} |\langle x, \xi \rangle|^n d\mu_P(x) < \infty.$$

Hence, all the polynomials in $\langle x, \xi \rangle$ belong to the Hilbert space $(L^2)_P = L^2(E^*, \mu_P)$.

Introduce the U-transform (see K. Saito and A. Tsoi [68]) defined by

$$U\varphi(\xi) = C_P(\xi)^{-1} \int_{E^*} e^{i\langle x, \xi \rangle} \varphi(x) d\mu_P(x), \quad \xi \in E,$$

for $\varphi \in (L^2)_P$.

Set

$$K(\xi, \eta) = \exp[\langle e^{i\xi} - 1, e^{i\eta} - 1 \rangle], \quad \xi, \eta \in E.$$

Then, the kernel $K(\xi, \eta)$ is proved to be positive definite, so that we can form a minimal reproducing kernel Hilbert space (RKHS) $\mathbf{F} = \mathbf{H}(K)$ with kernel K. (See Appendix 3.) Let the inner products in $(L^2)_P$ and $\mathbf{F}$ be denoted by $(\cdot, \cdot)$ and $(\cdot, \cdot)_{\mathbf{F}}$, respectively. Then, we have

$$(U\varphi, U\psi)_{\mathbf{F}} = (\varphi, \psi), \quad \varphi, \psi \in (L^2)_P.$$

As in the case of Gaussian white noise, we have the Fock space, that is a direct sum decomposition of $(L^2)_P$:

$$(L^2)_P = \bigoplus_{n=0}^{\infty} H_{P,n},$$

where $H_{p,n}$ is the space of the discrete chaos of degree n.

Under the U-transform the subspace $H_{p,n}$ is transformed to a subspace $\mathbf{F}_n$ of $\mathbf{F}$:

$$\mathbf{F}_n = \left\{ U\varphi(\xi) = \int_{R^n} F(u_1, u_2, \ldots, u_n) \prod_1^n P(\xi(u_j)) du^n, \quad \varphi \in H_{P,n} \right\},$$

where $P(\xi(u)) = e^{i\xi(u)} - 1$. The kernel F is a symmetric $L^2(R^n)$-function (that is $F \in \widehat{L^2(R^n)}$) and

$$\|\varphi\|^2 = n! \|F\|_{\mathbf{F}}^2.$$

Actually, the U-transform gives a bijection between $H_{P,n}$ and $\mathbf{F}_n$.

$$\mathbf{F}_n \cong \widehat{L^2(R^d)}. \tag{3.1.4}$$

In addition

$$\mathbf{F}_n \perp \mathbf{F}_m, \quad n \neq m. \tag{3.1.5}$$

Proposition 3.2 *The subspace $\mathbf{F}_n$ is spanned by*

$$\frac{\partial}{\partial Q(\eta_1(t_1))} \frac{\partial}{\partial Q(\eta_2(t_2))} \cdots \frac{\partial}{\partial P(\eta_n(t_n))} C_P(\xi),$$

where $Q(x) = e^{ix} - 1, \eta_i \in E$, and $t_i; i = 1, 2, \ldots, n$, are different.

In order to define *generalized Poisson noise functionals* we follow two steps. First the subspace $H_{p,n}$ is extended. Namely, kernel functions F can be taken to be a generalized function in the symmetric Sobolev space of order $-(n+1)/2$. Thus we have a larger space $H_{p,n}^{-n}$. The next step is to have a weighted sum of the H_n^{-n}: Take a decreasing sequence c_n of positive numbers to define

$$(L^2)_P^- = \bigoplus c_n H_{P,n}^{-n}. \tag{3.1.6}$$

This is the space of *generalized Poisson noise functionals.*

Accordingly in the RKHS $\mathbf{F}$, we have

$$\mathbf{F}^- = \bigoplus_n c_n \mathbf{F}_n^{(-n)}, \tag{3.1.7}$$

$$\mathbf{F}_n^{(-n)} \cong H_{P,n}^{-n}, \text{ under } U \text{ transform.}$$

Example 3.1 Remind that a stochastic bilinear form $\langle x, I(t) \rangle$ is a version of a Poisson process $P(t)$. The U-transform of a centered Poisson process is given by a kernel $\chi_{I(t)}(u) - \lambda t$. Hence the centered Poisson noise $\dot{P}(t) - \lambda$ corresponds to $\delta_t(u)$. Its p-th power has the kernel $\delta_t(u)^{p\otimes}$ and defines a generalized Poisson noise functionals; it is a polynomial in $\dot{P}(t)$.

Proposition 3.3

(i) $\frac{\partial}{\partial Q(\eta)}$ *is a derivation acting on $\mathbf{F}^-$. It is an annihilation operator and the domain involves $\mathbf{F}_n$.*

(ii) *The adjoint operator $(\frac{\partial}{\partial Q(\eta)})^*$ is defined and it is a creation operator.*

Compound Poisson process and compound Poisson noise (the time derivative of a compound Poisson process; it is often called a Lévy noise) shall be discussed later when we discuss innovation of a stochastic process or a random field. The case where the parameter space is taken to be $R^d, d \geq 1$, will also be discussed later.

Concerning a heuristic literature, the reader is recommended to follow N. Wiener's work on the discrete chaos [92].

We also see good examples in quantum optics (see e.g. [41]), where the intensity λ may be randomized.

The projection operator $E(t)$ in $(L^2)_P$ can be defined by

$$E(t)\varphi = E[\varphi|\mathbf{B}_t(P)],$$

where $\mathbf{B}_t(P)$ is the σ-field, with respect to which all the $\langle x, \chi_{I(s)}\rangle, s < t$, are measurable in the measure space $(E^*, \mathbf{B}, \mu_P)$.

Proposition 3.4 *$E(t)$ commutes with the projection $P_n : (L^2)_P \to H_{p,n}$.*

3.2 Functional equations for $C_P(\xi)$

(1) An equation characterizing holding time effect

We assume that $d = 1$.

Theorem 3.1 *A characterization of Poisson noise is given by the functional equation for the characteristic functional C_P:*

$$C_P(\xi) = \int_0^\infty \lambda \exp[-\lambda t + i\xi(t)] C_P(S_t\xi) dt, \tag{3.2.1}$$

where S_t is the shift operator:

$$(S_t\xi)(u) = \xi(u + t).$$

The converse is true under the assumption that C_P is a characteristic functional of a generalized process.

Proof. Let the jumping times be $t_0, t_1, \ldots, t_n$ such that $0 = t_0 < t_1 < \cdots < t_n = 1$. Set $\tau_i = t_i - t_{i-1}$, then $\{\tau_i\}$ are independent identically distributed exponential distribution with mean $\lambda^{-1}, \lambda > 0$.

Thus we can write

$$\dot{P}(t) = \sum \delta_{t_j},$$

and so we have

$$\begin{aligned}
C_P(\xi) &= E\left[\exp(i\langle \dot{P}, \xi\rangle)\right] = E\left[\exp\left(i\sum \langle \delta_{t_j}, \xi\rangle\right)\right] \\
&= E\left[\exp\left(i\sum_{j=1}^{\infty} \xi(t_j)\right)\right] = E\left[\exp\left(i\xi(t_1) + i\sum_{j=2}^{\infty} \xi(t_j)\right)\right] \\
&= E\left[\exp(i\xi(t_1)) \exp\left(i\sum_{j=2}^{\infty} \xi\left(t_1 + \sum_{k=2}^{j} \tau_k\right)\right)\right] \\
&= E\left[E\left(\exp(i\xi(t)) \exp\left(i\sum_{j=2}^{\infty} \xi\left(t + \sum_{k=2}^{j} \tau_k\right)\right) \Big| t_1 = t\right)\right] \\
&= E\left[\int \exp(i\xi(t)) \exp\left(i\sum_{2}^{\infty} \xi\left(t + \sum_{1}^{j-1} \tau_k\right) \lambda e^{-\lambda t}\right)\right] dt \\
&= \int \exp(i\xi(t)) \lambda e^{-\lambda t} E\left[\exp\left(i\sum \xi\left(t + \sum_{1}^{j-1} \tau_k\right)\right)\right] dt \\
&= \int \exp(i\xi(t)) \lambda e^{-\lambda t} C_P(S_t\xi) dt.
\end{aligned}$$

The proof of the converse is given by the following facts.

We know from the assumption of the continuity of $C_p(\xi)$ that there is a measure, in the Sobolev space E_{-1}, of order -1, which is defined by the $C_p(\xi)$. There exists a generalized stochastic process $\dot{Z}(t)$, the integral of which is $Z = Z(t) = \int_0^t \dot{z}(s)ds$.

The additivity of Z follows from the fact that $\dot{Z}(t)$ has independent values at every t. Let z be a sample function of Z. Then we see

$$\begin{aligned}
\langle \dot{z}(\cdot), S_t\xi(\cdot)\rangle &= \int \dot{z}(s)\xi(s+t)ds \\
&= \int \dot{z}(s-t)\xi(s)ds \\
&= \langle \dot{z}(\cdot - t), \xi\rangle.
\end{aligned}$$

The factor $e^{i\xi(t)}C_p(S_t\xi)$ which appears in (3.2.1) is understood to be the characteristic functional of $\delta_t(\cdot) + \dot{z}(\cdot - t)$, where $\delta_t(\cdot)$ and $\dot{z}(\cdot - t)$ are independent, t being fixed. Then the integral in (3.2.1) shows that it is the expectation with respect to the randomized t subject to the exponential distribution.

Thus the characteristic functional in question is expressed in the form

$$E\left[e^{i(\sum_j \langle \delta_{t_j}, \xi_j \rangle)}\right]$$

which is in agreement with the characteristic functional of Poisson noise.

(2) The case of random intensity

The Poisson noise with randomized intensity plays an important role in quantum optics (see [41], Chapter 2). It is also characterized by the functional equation for the characteristic functional. Assume that the intensity is subject to the exponential distribution with parameter α and let $\widehat{C}_P(\alpha, \xi)$ denote the Laplace transform of $C_P = C_P^\lambda, \lambda \geq 0$. Then, the mean of C_P^λ with randomized λ is $\alpha\, \hat{C}_P(\alpha, \xi)$. A characterization in question comes from

$$\widehat{C}_P(\alpha, \xi) = -\frac{d}{d\alpha} \int e^{i\xi(t)} \widehat{C_P}(\alpha + t, S_t \xi) dt.$$

Further properties of a Poisson noise will be discussed in Section 6.5.

3.3 Observation of 1-dimensional parameter Poisson noise

In what follows is the definition of Poisson noise, and its characteristic properties will be discussed with special emphasis on the optimality. This will help us to give an effective determination of Poisson noise with one or higher dimensional parameter.

A. Characteristic functional

A Poisson noise is the time derivative of Poisson process $P(t)$ and is denoted by $\dot{P}(t)$, as before.

As is well known, $\dot{P}(t)$ is a *stationary* generalized stochastic process with independent values at every instant t.

The time parameter is now restricted to a finite interval, say $I = [0, 1]$. To have the corresponding characteristic functional, it is necessary to take the basic nuclear space to be a subspace of $L^2(I)$.

In this section, the jump points of $P(t), t \in I$, will be denoted by $\tau_1, \tau_2, \ldots$, successively in increasing order.

Remark 3.1 *The continuous bilinear form $\langle \dot{P}, \xi \rangle$ extends to a random variable $\langle \dot{P}, f \rangle$, so as to be linear in $f \in L^2([0,1])$, and has the probability distribution determined by the characteristic function*

$$\exp\left[\lambda \int_I (e^{izf(t)} - 1)dt\right] \equiv C_P^I(zf), \quad z \in R.$$

If f is taken to be an indicator function $\chi_{[0,t]}$, then the random variable obtained as above is in agreement with $P(t)$.

B. Some conditional probabilities

Our study will be concentrated on Poisson noise over a unit time interval I. Consider the event A_n on which $P(1,\omega) = n$, where n is any non-negative integer. Then, the collection $\{A_n, n \geq 0,\}$ is a partition of the entire ω-set Ω. Namely, up to a set of measure 0, the following relations hold:

$$A_n \cap A_m = \phi, \quad n \neq m; \quad \bigcup A_n = \Omega.$$

Set

$$X_j = \tau_j - \tau_{j-1}, \quad j = 1, 2, \ldots, n \quad (\tau_0 = 0, \tau_{n+1} = 1) \tag{3.3.1}$$

on A_n.

Proposition 3.5 *Under the assumption $P(1) = n$, the probability distribution of the random vector $(X_1, X_2, ..., X_{n+1})$ is uniform on the simplex $\sum_{j=1}^{n+1} x_j = 1, x_j \geq 0$.*

Proof. It is known that the time intervals holding a constant value of a Poisson process $P(t), t \geq 0$, are all subject to an exponential distribution with density function $\lambda \exp[-\lambda t], \lambda > 0, t \geq 0$, and they are independent. Hence, the joint distribution of holding times is subject to the direct product of exponential distribution. Namely,

$$P(X_j \geq t_j, \ 1 \leq j \leq n+1) = \exp\left(-\lambda \sum_1^{n+1} t_j\right).$$

It can easily be seen that the density function of the conditional joint distribution density is constant over the symplex $\sum_1^{n+1} t_j = 1$. Thus the assertion is proved.

In this sense, Poisson process enjoys a sort of *optimality* (maximum entropy), since the uniform distribution is involved.

Corollary 3.1 *The probability distribution function of each X_j is $1 - (1-x)^n, 0 \leq x \leq 1$.*

C. Conditional characteristic functionals

Let the time parameter be restricted to $I = [0,1]$, again. Take the event A_n given before. The *conditional characteristic functional* $C^I_{P,n}(\xi)$ is defined as follows:

$$C^I_{P,n}(\xi) = E[e^{i\langle \dot{P},\xi\rangle}|A_n].$$

Proposition 3.6 ([81]) *We have*

$$C^I_{P,n}(\xi) = \left(\int_0^1 e^{i\xi(t)}dt\right)^n. \tag{3.3.2}$$

Proof. The proof proceeds by induction. Assume that the assertion is true for n. Then, we have

$$\begin{aligned} E\left[e^{i\sum_1^{n+1}\xi(\tau_j)}\Big|A_n\right] &= E_{\tau_{n+1}}\left[E\left[e^{i\sum_1^{n+1}\xi(\tau_j)}\right]\Big|\tau_{n+1} = x\right] \\ &= \int_I (n+1)(1-x)^n e^{i\xi(x)} E\left[e^{i\sum_1^n \xi(\tau_j)}\Big|P(1-x) = n\right] dx \\ &= \int_I (n+1)(1-x)^n e^{i\xi(x)} \left(\frac{1}{1-x}\int_0^{1-x} e^{i\xi(t)}dt\right)^n dx \\ &= \left(\int_I e^{i\xi(t)}dt\right)^{n+1}. \end{aligned}$$

Thus, the assertion is proved.

Remark 3.2 *Noting that $P(A_n) = \frac{\lambda^n}{n!}e^{-\lambda}$, it is proved that*

$$\begin{aligned} \sum_0^\infty C^I_{P,n}(\xi)\frac{\lambda^n}{n!}e^{-\lambda} &= \sum_0^\infty \left(\int_I e^{i\xi(t)}dt\right)^n \frac{\lambda^n}{n!}e^{-\lambda} \\ &= C^I_P(\xi), \end{aligned}$$

which is in agreement with the characteristic functional of Poisson noise without any restriction on its value.

3.4 Construction of 1-dimensional Poisson noise

Let a collection $\{A'_n, n \geq 0\}$ be any partition of the entire sample space Ω such that $P(A'_n) = \frac{\lambda^n}{n!}e^{-\lambda}$.

Let $Y_k^{(n)}$, $1 \leq k \leq n$, be a sequence of independent identically distributed random variables, on the probability space $(A'_n, P(\cdot|A'_n))$, which are distributed uniformly on $I = [0,1]$. The order statistics of $Y_k^{(n)}$ gives us an ordered sequence $Y_0^{(n)} \le Y_{\pi(1)}^{(n)} \le Y_{\pi(2)}^{(n)} \le \cdots \le Y_{\pi(n)}^{(n)} \le Y_{\pi(n+1)}^{(n)}$, where $Y_0^{(n)} = 0$ and $Y_{\pi(n+1)}^{(n)} = 1$. Then, the probability distribution of $Y_{\pi(k+1)}^{(n)} - Y_{\pi(k)}^{(n)}$ has the common density function $n(1-x)^{n-1}$, for every $0 \le k \le n$. This fact can be proved by mathematical induction.

Theorem 3.2 *The random vector* $\{Y_{\pi(k+1)}^{(n)} - Y_{\pi(k)}^{(n)},\ 1 \le k \le n\}$ *on* A_n' *has the same distribution as that of* $\{X_k,\ 1 \le k \le n\}$, *given in* (3.3.1).

This theorem suggests to us how to construct a Poisson noise on I. In other words, with these $Y_k^{(n)}$'s we can form a noise $V^{(n)}(t,\omega)$ in such a way that

$$V^{(n)}(t,\omega) = \sum_{k=1}^{n} \delta_{Y_k(\omega)}^{(n)}(t), \quad \omega \in A_n. \tag{3.4.1}$$

Corollary 3.2 *The generalized process* $V^{(n)}(t,\omega)$, $t \in I$, *is the same process as a Poisson noise* $\dot{P}(t,\omega)$, $t \in I$, *on the conditional probability space* $(A_n, P(\cdot|A_n))$, *defined before.*

We now have

Theorem 3.3 *Let a system* $\{Y_j^{(n)}(\omega)\}$ *be independent uniformly distributed random variables on* I *with* $\omega \in A_n$. *By arranging the* $\{Y_j^{(n)}(\omega)\}$ *in increasing order, we have* $V^{(n)}(t,\omega)$, *by* (3.4.1). *Set*

$$V(t,\omega) = V^{(n)}(t,\omega), \quad \omega \in A_n'; \ n = 0, 1, \ldots \tag{3.4.2}$$

on I. *Then,* $V(t,\omega), \omega \in \Omega$, *is a Poisson noise with the parameter* t *running through* I.

(Cf. [46])

3.5 Construction of d-dimensional parameter Poisson noise

Having been suggested by the observation in Section 3.4, we define the R^d-parameter Poisson noise in this section.

In addition, we focus our attention to the optimality properties in terms of entropy. The maximum entropy is obtained since Poisson noise is formed by independent and uniformly distributed random functions which is taken to be the delta functions that corresponds to jump points of Poisson process.

In the one dimensional parameter case the jump points are naturally ordered on the real line. However, there arises a question on how to define a Poisson noise by ordering the jump points suitably in the two dimensional region as well as in higher dimensional parameter case. This leads us to think of the idea and thus we give the effective definition of the higher dimensional Poisson noise as a generalization of 1-dimensional parameter case.

We have discussed the conditional characteristic functional of Poisson noise when the number of jumps is given in a specified time interval in one dimensional parameter case, in the previous section. Its generalization to a higher dimensional parameter case can be done in the same manner discussed above.

The probability distribution of the random vectors, indexed by a point in I^d, that is Poisson noise with d-dimensional parameter, is invariant under the permutation of the coordinates in I^d, that is, under the symmetric group.

A. Effective determination of R^d-parameter Poisson noise

Our question is how to define the R^d-parameter Poisson noise. By refering the characteristic properties of the one-dimensional parameter case, in particular optimality, we can give a definition of the noise in question as a multi-dimensional generalization of what we have observed in Section 3.3. A sample function should now be a collection of delta-functions which are randomly scattering.

Having limited the parameter set, say a unit d-dimensional cube $D = [0,1]^d$, and fixing the event A_n on which there are delta functions as many as n.

Let a collection $\{A_n, n \geq 0\}$ be any partition of the entire sample space Ω such that $P(A_n) = \frac{\lambda^n}{n!}e^{-\lambda}$.

Now we define a D-parameter Poisson noise as follows.

Definition 3.1 Let the random vectors

$$Y_j^{(n)}(t,\omega) = (Y_{j1}^{(n)}(t,\omega), \ldots, Y_{jd}^{(n)}(t,\omega)); \quad j = 1, \ldots, n$$

be such that the components are independent and uniformly distributed random variables on $(A_n, P(\cdot|A_n))$, for $t \in D$. Set

$$V^{(n)}(t,\omega) = \sum_{i=1}^{n} \delta_{Y_i^{(n)}(\omega)}(t) \quad \text{on } A_n. \tag{3.5.1}$$

A *D-parameter Poisson noise* is defined by

$$V^D(t,\omega) = V^{(n)}(t,\omega), \quad \omega \in A_n; \ n = 0, 1, \ldots; \ t \in D. \tag{3.5.2}$$

To determine the probability distribution, the conditional characteristic functional is given in the following proposition. We now see that the event A_n defined above is the event that n delta functions are sitting in D.

Proposition 3.7 *Given the event A_n. The conditional characteristic functional $C_{P,n}(\xi) = E[e^{i\langle V,\xi\rangle}|A_n]$ of a D-parameter Poisson noise is of the form*

$$C_{P,n}^D(\xi) = \left(\int_D e^{i\xi(t)}dt^d\right)^n. \tag{3.5.3}$$

Proof is similar to the one-dimensional parameter case (in Section 3.3), so is omitted.

The (unconditional) characteristic functional of a Poisson noise is computed as follows. Take the Poisson distribution with intensity λ to have an average. Then,

$$\begin{aligned}\sum_0^\infty C_{P,n}^D(\xi)P(A_n) &= \sum_0^\infty C_{P,n}^D(\xi)\frac{\lambda^n}{n!}e^{-\lambda}\\ &= \exp\left(\lambda\int_D (e^{i\xi(t)}-1)dt^d\right)\end{aligned}$$

which is the same form as the characteristic functional of one dimensional parameter Poisson noise expressed in (3.1.1).

Now take D to be a disc in R^d with center O (origin). Then, the characteristic functional $C_{P,n}^D(\xi)$ is obviously invariant under rotations around O. This means that n delta functions of R^d-parameter Poisson noise is invariant under the $SO(d)$. Namely, the distribution is symmetric with respect to the rotations.

At this stage, it is not hard to define the conditional characteristic functional with R^d-parameter case, just by replacing D with R^d. The $C_{P,n}^{R^d}(\xi)$ is simply written as $C_{P,n}(\xi)$.

Proposition 3.8 *The weighted sum of $C_{P,n}(\xi)$ with weight $\frac{\lambda^n}{n!}e^{-\lambda}$ is expressed in the form*

$$C_P(\xi) = \exp\left(\lambda\int_{R^d}(e^{i\xi(t)}-1)dt^d\right). \tag{3.5.4}$$

Again take D to be I^d. We use the usual notation Z for the set of integers in the following. Let $D+m = \{t+m;\ t\in D\}$, $m\in Z^d$ and define

$$V^m(t,\omega) = V^{D+m}(t,\omega), \tag{3.5.5}$$

in the same manner as $V^D(t,\omega)$ so as $\{V^m(t,\omega), m \in Z^d\}$ to be independent. A construction of a Poisson noise with the characteristic functional (3.5.4) is given by

Theorem 3.4 *The characteristic function of*

$$V(t) = \sum_{m \in Z^d} V^m(t,\omega) \tag{3.5.6}$$

is $C_P(\xi)$, given by (3.5.4).

Proof is obvious.

Thus, the $V(t)$, given by (3.5.6), is an R^d *parameter Poisson noise.*

There remains a question on the reason why we take an average by using the Poisson distribution. An elementary and plausible interpretation to take such a weight of a Poisson distribution is given as follows.

We are suggested to take the weight as is familiar in the partition function in statistical mechanics. For the ideal gas, the energy at the level (n_x, n_y, n_z) is denoted by $\varepsilon(n_x, n_y, n_z)$ and the partition function U_n for n particles is given by

$$U_n = \frac{1}{n!} e^{-c\sum_{j=1}^n \varepsilon(n_{x_j}, n_{y_j}, n_{z_j})}, \quad c : \text{constant}.$$

Since each particle has the same energy, we must have

$$U_n = \frac{1}{U} \frac{\lambda^n}{n!}.$$

With the choice of U so as to be $\sum_0^\infty U_n = 1$, we are given Poisson distribution with intensity λ proportional to the energy level.

B. Restriction of parameter

Consider a d-dimensional parameter Poisson noise. Let n be fixed and let the event A_n be given. For each $\omega \in A_n$ there exist delta functions as many as n, the coordinates of which are d-dimensional vectors $Y_1^d, Y_2^d, \ldots, Y_n^d$. Take $d' < d$. By the orthogonal projection

$$\pi(d, d') : D = [0,1]^d \to D' = [0,1]^{d'},$$

we are given d-dimensional vectors

$$Y_k^{d'} = \pi(d, d') Y_k^d, \quad 1 \le k \le n.$$

Note that $Y_k^{d'}$'s are different almost surely, since Y_k^d's are independent and uniformly distributed.

With the help of the $Y_k^{d'}$'s we can define d'-dimensional parameter Poisson noise. The conditional characteristic functional of $D' = [0,1]^{d'}$ parameter Poisson noise is obtained as

$$C_{P,n}^{D'}(\xi) = \left(\int_{D'} e^{i\xi(t)} dt^{d'} \right)^n \tag{3.5.7}$$

and consequently, the (unconditional) characteristic functional of a Poisson noise can be computed as is shown in Section 3.3. By reminding the Poisson distribution with intensity λ, the characteristic functional is obtained as

$$\begin{aligned} C_{P,D'}(\xi) &= \sum_0^\infty C_{P,n}^{D'}(\xi) \frac{\lambda^n}{n!} e^{-\lambda} \\ &= \exp\left(\lambda \int_{D'} (e^{i\xi(t)} - 1) dt^{d'} \right) \end{aligned}$$

which is in the same form as the characteristic functional of one dimensional parameter Poisson noise expressed in (3.1.1), with time domain D'.

Such a method of parameter restriction together with the results will be used to have the variation of random field $X(C)$ formed by a Poisson noise.

C. Poisson noise parameterized by a manifold

We are interested in an effective determination of Poisson noise parameterized by a manifold, in particular by a sphere S^d.

Let d be fixed, and let Y_k, $1 \leq k \leq n$, be a sequence of independent random variables taking values in S^d. Following the idea of determination of a Poisson noise with optimality in mind, we assume that Y_k's are distributed uniformly on S^d. Denote by σ the uniform probability distribution on S^d. Then, the characteristic functional is

$$\begin{aligned} C_{S^d,n}(\xi) &= \int_{S^{d \times n}} e^{i \sum_1^n \xi(t_k)} d\sigma(t_1) \cdots d\sigma(t_n) \\ &= \left(\int_{S^d} e^{i\xi(t)} d\sigma(t) \right)^n . \end{aligned}$$

This is thought of as the conditional characteristic functional for given n. The unconditional characteristic functional is to be

$$\begin{aligned} C_{S^d}(\xi) &= \sum_0^\infty C_{S^d,n}(\xi) \, \frac{\lambda^n}{n!} e^{-\lambda} \\ &= \exp\left(\lambda \int_{S^d} (e^{i\xi(t)} - 1) d\sigma(t) \right) . \end{aligned}$$

This expression will be used when the variation of a random field $X(C)$ is actually computed, where the $X(C)$ is a functional of Poisson noise $V(u)$, $u \in (C)$.

3.6 Invariance of Poisson noise

It is obvious that the time shift leaves the probability distribution μ_p of a Poisson noise invariant, since

$$C_P(S_t^k \xi) = C_P(\xi), \quad (S_t^k \xi)(u) = \xi(u - te_k), \ \ u \in R^d, \tag{3.6.1}$$

where e_k is the unit vector in the k-th direction of R^d.

Obviously, orthogonal group $O(d)$ acting on the parameter space R^d presents invariance of Poisson measure.

Proposition 3.9 *The probability distribution μ_P of an R^d-parameter Poisson noise is invariant under the actions*

(1) *the orthogonal group $O(d)$, and*
(2) *the shifts of R^d.*

Interesting property can be seen in the dilation of R^d-vectors:

$$\tau_t : u \to ue^{at}, \quad u \in R^d, \ \ a > 0.$$

Then the characteristic functional $C_P(\xi)$ changes into

$$\exp\left[\lambda \int (e^{i\xi(ue^{at})} - 1)du^d\right] = \exp\left[\lambda e^{-dat} \int (e^{i\xi(v)} - 1)dv^d\right].$$

Namely, the intensity changes from λ to λe^{-dat}, but distribution remains to be of Poisson type.

Some invariance of the probability distribution of the conditional Poisson noises have already been discussed in Section 3.5.

Before functionals of Poisson noise and their analysis are discussed, it is worth mentioning some properties related to the random placement of n points. Given equally distributed n delta functions in $[0, 1]$ or equivalently, given an event $A_n(C\ \Omega(P))$ on which the probability concerning the positions of n delta functions is invariant under the permutation of those delta functions. More rigorously, the positions of the delta functions are represented by the random variables (random vectors in the R^d-parameter case) $X_1, X_2, \ldots, X_n$ which are distributed independently and identically

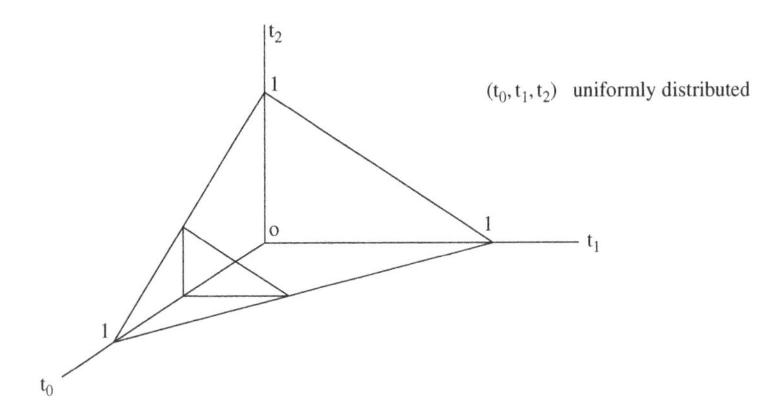

Fig. 4. Simplex.

on [0, 1]. Obviously the probability distribution of the random vectors is invariant under the action of the permutation group $S(n)$. Taking the order statistics of X_j's we come to the same situation as in Section 3.4.

The placement or the arrangement of n points discussed above can be realized by a Poisson noise (or a Poisson process) as follows. Now D is taken to be the unit interval. Let $V(t), t \in D$, be a Poisson noise. Its probability distribution μ_P^D is given on the space $E(D)^*$, which is the dual space of a nuclear space $E(D)$ dense in $L^2(D)$. The event A_n corresponds to a subset $B(n)$ of $E(D)^*$ by the measurable mapping:

$$\omega \rightarrow x \in E(D)^*, \quad \omega \in A_n,$$

where almost all x (with respect to the measure μ_P) is viewed as a sample $V(\cdot, \omega)$ of the Poisson noise for $w \in A_n$. Such a transformation can be done for every n and defines a mapping from Ω into $E(D)^*$. Under this extended mapping the group $S(n)$ naturally defines a transformation g acting on $E(D)^*$ which leaves the measure μ_P invariant.

3.7 Multi-dimensional parameter Poisson sheet

The characteristic functional $C_P(\xi)$ of a Poisson noise with a given parameter set has the same form, regardless the domain of the integral, in expression as was seen before. The Poisson sheet indexed by a rectangular domain is useful to observe the properties of a Poisson noise in this case, too. Actually a review and a supplement will be given in Section 7.7. The R^d-parameter Poisson sheet $P^d(t), t = (t_1, t_2, \ldots, t_d)$, is defined by a stochastic linear functional of R^d-parameter Poisson noise $\dot{P}^d(t)$ in such a way that

$$P^d(t) = \langle \dot{P}^d, \chi_{I(t)} \rangle, \quad I(t) = \prod_{j=1}^{d} [0, t_j].$$

It has independent variations and the probability distribution is of Lattice type.

3.8 Compound Poisson noise

A compound Poisson noise appears under mild assumptions in the general innovation, for which the Lévy decomposition is applicable. Hence, many interesting properties are obtained by reducing the given compound Poisson noise to individual elemental Poisson noise, which is the component of the innovation of the random complex system in question.

The Lévy decomposition of the innovation in the standard case is explained as follows. To make the matters somewhat simpler, we still take $d = 1$. The innovation is a system of idealized elemental random variables, so that it is reasonable to assume that it is a time derivative of an additive process which is continuous in probability and hence, we have a Lévy process $L(t)$. Although there is some restriction, we may consider that the increment of the additive process is stationary.

Under these assumptions, the innovation, which is now denoted by $\dot{L}(t)$, has the characteristic functional

$$C_L(\xi) = \exp\left[-\frac{\sigma^2}{2} \|\xi\|^2 + \int \int (e^{i\xi(t)u} - 1 - i\xi(t)u) dn(u) \right], \qquad (3.8.1)$$

where $dn(u)$ is the Lévy measure satisfying the condition

$$\int \frac{u^2}{1+u^2} dn(u) < \infty. \qquad (3.8.2)$$

If the Gaussian part $-\frac{\sigma^2}{2}$ is missing, we are given a compound Poisson noise.

The notion of the innovation will be discussed in details in Chapter 8, but here we have mentioned an idea briefly.

From the expression of $C_L(\xi)$, we understand $L(t)$ is a sum of Brownian motion (up to constant) and a superposition of independent Poisson process with different jumps.

The reduction to an elemental Poisson noise is done in the standard manner, although computability (or possibility of measurement) problem is involved. While, one must meet a difficulty in constructing the superposition of continuously many Poisson noises, which are mutually independent. In reality, the quasi-convergence of the sum (the integral) of them is the basic method to have a compound Poisson noise. (See e.g. P. Lévy [46].)

3.9 The space (P)

As is mentioned in the last section, it is an interesting and in fact important problem to have the decomposition of a given compound Poisson process into individual elemental Poisson processes with different jumps. Conversely, the superposition of those elemental Poisson processes is another profound problem. To deal with those questions it is necessary to provide a suitable space of random variables and of stochastic processes.

Also, having motivated by many problems in applications and reminded our original idea of innovation theory, we are going to introduce a new class of functionals of white noise either of Gaussian or of Poisson type. We also try to find available tools for the new analysis on the space $(\mathbf{P})$.

Take a probability space $(\Omega, \mathbf{B}, P)$ on which a vector space $(\mathbf{P})$ of complex-valued random variables is given. The topology which is to be introduced to this space is defined by the convergence in probability. It is often useful to take a metric $d(X, Y)$ for $X, Y \in (\mathbf{P})$, defined by

$$d(X,Y) = E\left(\frac{|X-Y|}{1+|X-Y|}\right).$$

This metric, as is well known, defines the same topology as that defined by convergence in probability.

The following assertion is easily proved, since the mean square convergence defines stronger topology than that defined by convergence in probability.

Proposition 3.10 *The inclusions*

$$(L^2) \subset (\mathbf{P}), \quad (L_P^2) \subset (\mathbf{P})$$

hold and give us continuous injections.

To make the discussion to be more concrete, we sometimes specify the space $(\mathbf{P})$ to be a collection of random variables which are measurable with respect to a specified $\mathbf{B}$, say $\mathbf{B} = \mathbf{B}(\mathbf{X}), \mathbf{X} = \{X(C), C \in \mathbf{C}\}$.

Before we study the analysis on the space (**P**) it seems helpful to remind useful applications what we have experienced before.

The first is that we have often met problems for which we cannot assume the existence of their second order moments of random variables involved there. As we shall see later in Chapter 8, basic components of the innovation are Gaussian noise and (compound) Poisson noise. In order to decompose the innovation into elemental noise, we are suggested to appeal to the Lévy–Itô decomposition of a Lévy process. For this purpose it is necessary to observe individual sample functions of a Brownian motion and those of a compound Poisson process. The trick like jump-finding can be controlled within the framework of neither (L^2) nor $(L^2)_P$. This remark might be thought of a fact which is outside of the matter to be attentioned, but not quite. As soon as we come to computability problem in application, we meet this problem.

The next example is a method of subordination of a stochastic process. Roughly speaking it is a random time change of a stochastic process. We know interesting examples in applications like astrophysics data processing, where a mathematical model is formed by subordination applied to a standard stochastic process.

The subordination is interesting in itself, and in addition, it provides a profound operation in functional analysis, being closely related to stochastic analysis. It is understood to be a kind of analysis outside of (L^2) theory.

Another interesting topic is concerned with optimality in terms of sample functions. For a Brownian motion, its sample function is continuous and very irregular as is investigated by many ways. It is also interesting to note intuitively, each sample function, viewed as a trajectory of a Brownian particle, travels so as to occupy a territory efficiently. As for a Poisson noise we observed a different kind of optimality in Section 3.3.

Actions, defined by rotations of E, on individual sample function of a stochastic process or a random field are often efficiently dealt within the space (**P**). A good example is the projective invariance of white noise which is explained by the normalized Brownian bridge. It is known that the probability distribution of the normalized Brownian bridge is invariant under the projective transformations of the time parameter. We can say more precisely the invariance in terms of sample function property in (**P**).

Now, we can come to the case of random fields. The reversibility of certain process and field of white noise functionals can be shown. Also, the reversibility can often be found in Poisson case.

The analysis of (compound) Poisson functionals, using various operations acting on them can be dealt within this framework without much restrictions also in (**P**).

A stochastic process $X(t)$ and random field $X(C)$ discussed in this section are therefore assumed to live in (**P**) and continuous. The most powerful tool to analyze them would be the characteristic functional.

(i) For a continuous stochastic process $X(t)$, we have the characteristic functional

$$C_X(\xi) = E(e^{i\langle X,\xi\rangle}), \quad \xi \in E, \langle X,\xi\rangle = \int X(t)\xi(t)dt.$$

For the case of R^d-parameter random fields, we have the same expression as above for characteristic functionals.

(ii) Consider a random field $X(C)$, C being an ovaloid in the plane. The parameter C is represented by a pair of functions, say $(a_C(t), b_C(t))$, and the discussion is reduced to the case (i) with parameter C.

The Hopf equation is a good example for which the characteristic functional method can be applied efficiently. (See Section 7.8 and [39].)

We take a random field $X(C) = X(C,\omega), C \in \mathbf{C}, \omega \in \Omega(P)$, and introduce a space (**P**) of random variables which are functions of $X(C)$'s. Then, like $X(t), t \in R^d$, or $X(f)$, parameterized by a function f, can be discussed similarly, even easily.

Let $\mathbf{B}(X)$ be the smallest sigma-field of subsets of Ω, with respect to which all the $X(C)$'s are measurable. Let $\mathbf{C}$ be topologized by the Euclidean metric ρ, namely the distance $\rho(C, C')$ is defined by

$$\rho(C, C') = \max_{u\in C, u'\in C'} \rho(u, u'),$$

with respect to this metric, we assume that $X(C)$ is continuous in probability.

Given such a random field $X(C)$. We introduce a space (**P**) or more precisely denoted by $(\mathbf{P})(X)$, which involves $\mathbf{B}(X)$-measurable, complex valued random variables. We introduce the *limit in probability* topology to the space (**P**). The $X(C)$ is now viewed as a continuous curve in (**P**) with parameter C.

Limit theorems for $X(C)$ are discussed within the framework of (**P**). Our most important direction is the variational calculus. This can be established in the following manner. To avoid non-essential complex description, we assume that $d = 2$. By assumption C can be represented by a smooth, univalent vector valued function, say $f(t) = (f_1(t), f_2(t))$, $t \in [0, 1]$.

We can therefore define a characteristic functional $C_X(\xi)$, $\xi = (\xi_1, \xi_2) \in E \times E$. Hence, the reproducing kernel Hilbert space $\mathbf{H}(C)$ can be defined (see Appendix 3) so that Hilbert space technique can be used with visualized expressions.

On the other hand, still in line with the characteristic functional method, we are given the $\mathcal{T}$-transform of $X(C) = X(f)$, to have a functional $U(\xi, f), \xi \in E \times E$, depending on f. Hence, the theory of (non-random) functional analysis is ready to be applied. Actual computation of the variational calculus will be discussed in Chapter 7.

Chapter 4

Random Fields

This chapter is devoted to a brief introduction to the general theory of random fields, which are considered as random evolutional complex systems that develop as a space-time (or more general) parameter varies. One can compare this theory with that for ordinary stochastic processes. It is hard to imagine how the evolution, which is described by a variation of the random field, presents extremely complex behaviour and shows much profound probabilistic properties. Concrete and detailed properties will be shown in the following chapters, but a quick overview will be given in this chapter.

4.1 Processes and fields as white noise functionals

Some of the foundation of white noise analysis and the basic notions of random fields as well as a variational calculus, which is the basic tool of our study, will be provided in this chapter. A white noise $W^d(u) = W^d$ (often d is omitted if no confusion occurs), parameterized by $u \in R^d$ is a system of idealized elemental random variable which is stationary in u and is standard Gaussian in distribution, being introduced in the space E^* of generalized functions on R^d. Recall that its characteristic functional is expressed in the form

$$\begin{aligned} C(\xi) &= \int_{E^*} \exp[i\langle x, \xi\rangle] d\mu(x) \\ &= \exp\left[-\frac{1}{2}\|\xi\|^2\right], \end{aligned} \tag{4.1.1}$$

where ξ is a member of a nuclear space E, dense in $L^2(R^d)$.

The measure space (E^*, μ) is a *white noise* with R^d parameter and is a rigorous realization of the $W(u)$. The probability distribution μ is called the *white noise measure*. Thus, we are given the complex Hilbert space $(L^2) = L^2(E^*, \mu)$, a member of which is a functional of white noise, or simply called *white noise functional*, expressed in the form like $\varphi(x)$, where x is a member of E^* with measure μ. Here, μ-almost all x are sample functions of W.

Needless to say, in the case $d = 1$, W^1 is nothing but $\dot{B}$.

We shall discuss, in this chapter, functionals, in general nonlinear functionals of white noise W^d, which is often written simply by W or by x to make the notation simple.

4.2 Random fields $X(a)$ and $X(f)$

(1) Higher dimensional parameter case

A stochastic process is a random function $X(t)$ indexed by the time parameter t and it gains, at each instant t, a new information independent of the past values, where t is tacitly assumed to be the present time. One of the generalizations of this notion can be given by a random field $X(a)$, where a is a vector running through a domain of higher dimensional Euclidean space or a Riemannian manifold.

We shall begin with a field $X(a), a \in R^d$.

Most important example is the Lévy Brownian motion $X(a) = B(a)$, $a \in R^d$, the definition of which is given in Section 1.2. This is a natural generalization of the ordinary Brownian motion $B(t), t \in R$.

Many interesting results have been obtained starting from the book by P. Lévy 1948 [46]; there are basic results, due to McKean [56] and Chentsov [4] for the representations. Mckean's expansion of $X(a)$ is most significant in our approach.

One is now interested in a generalization of the relation between Brownian motion $B(t)$ and white noise $W(t)$. In one dimensional parameter case, it is quite simple; $W(t) = \dot{B}(t)$. However, in the higher dimensional parameter case, the situation is quite different. We therefore wish to discover principles behind this matter. To be concretized, it would be fine if we can find a representation of Brownian motion in terms of white noise.

H.P. McKean and N.N. Chentsov gave explicit form of the representation; both are essentially the same. McKean defined the Lévy Brownian

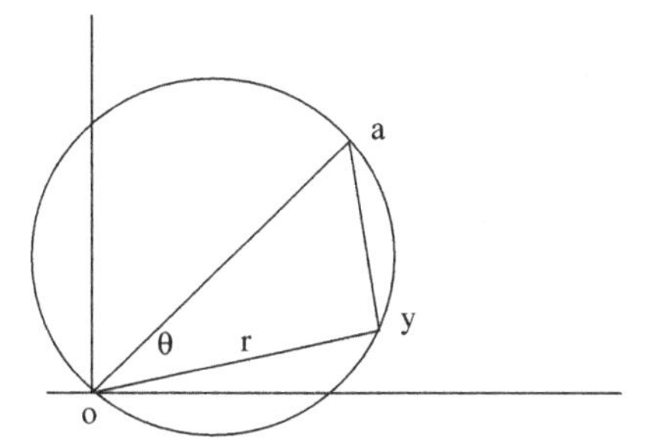

Fig. 5. White noise → Lévy Brownian motion (see McKean [56]).

motion as

$$X(a) = c(d) \int\int_{0 \le r \le (\overline{oa}, y)} w(r, \theta) dr\, d\theta, \quad y = y(\theta) \in S^{d-1}, \tag{4.2.1}$$

where $(\cdot, \cdot)$ is the inner product in R^d, o is the origin of R^d, $w(r, \theta)$ is a white noise on $(0, \infty) \times S^{d-1}$, $d\theta$ being the surface element of S^{d-1} and

$$c(d)^2 = 2(d-1) \int_0^{\frac{\pi}{2}} \sin^{d-2}\theta\, d\theta. \tag{4.2.2}$$

Proposition 4.1 *The system $\{X(a)\}$ is a Lévy Brownian motion.*

For the proof, we can verify the covariance to be $\Gamma(a, b)$ given in Section 1.2.

Example 4.1 Consider the Lévy Brownian motion in 2-dimensional case. The formula (4.2.1) gives us

$$X(a) = c(2) \int\int_S w(r, \theta) dr\, d\theta,$$

where S denotes the circle with diameter $\overline{oa}$.

We can easily see that

$$E[X(a)^2] = \rho(o, a).$$

Thus, $X(a)$ is a Lévy Brownian motion. (See Section 1.2.)

Remark 4.1 *For higher dimensional parameter case, such a representation can be obtained similarly as above.*

Lévy's Brownian motion is indeed a good representative of random fields with multi-dimensional parameter in many respects. Not only the Brownian motion itself but also derived fields and processes are discussed with much interest. For instance, the results by one of the authors illustrate a property of its complex dependence (Si Si [70]).

We need a system of innovations in many directions; either along the radial directions, or along the smooth curves, or ordered surfaces, and so forth. Hence, the $X(a)$ has infinite multiplicity along the direction of development of a. This is another example of significant characteristics of the Brownian motion $X(a)$ with multi-dimensional parameter.

Remind the idea of generalizing the innovation or the definition of a stochastic process. As an immediate consequence, we can consider simple as well as multiple Markov properties.

It is noted that to get a system of the innovations of the $X(a)$, we need higher order differential operators in the variable r in the case of odd d, while if d is even, a fractional order variational operators are required. This fact reflects the Markov properties of $X(a)$, as it were, of fractional order.

(2) Euclidian free field

Given a functional

$$C_f(\xi) = \exp\left[-\frac{1}{2}\int \frac{|\hat{\xi}(\lambda)|^2}{\lambda^2 + m^2} d\lambda\right], \quad \xi \in E \subset L^2(R^d), \tag{4.2.3}$$

where $\hat{\xi}$ is the Fourier transform of ξ and $m > 0$ is the mass.

It is easy to see that C_f is written as

$$C_f(\xi) = \exp\left[-\frac{1}{2}\langle \xi, \Gamma\xi\rangle\right], \tag{4.2.4}$$

and the operator Γ is $(-\Delta + m^2)^{-1}$, for which $\langle \xi, C\xi\rangle$ is a positive continuous nondegenerate bilinear form on $E \times E$.

The $C_f(\xi)$ is a characteristic functional, so that there exists a probability measure $d\mu_f$ on E^*. In terms of quantum dynamics, the measure space $(E^*, d\mu_f)$ is the Euclidean free field.

It is a Gaussian measure space with mean zero and covariance functional $\langle \xi, \Gamma\xi\rangle$. Hence, there is a Gaussian random field $X(\xi)$ with

$$E[X(\xi)] = 0, \quad E[X(\xi)X(\eta)] = \langle \xi, \Gamma\eta\rangle = \langle \Gamma\xi, \eta\rangle.$$

More general Euclidean fields together with the relation with Bosonic quantum fields will be discussed in Section 5.6.

(3) Random fields $\boldsymbol{X}(f)$

First we consider the case where f runs through some function space, say $L^2(R)$. Or more specifically, assume that f is a vector in the space l^2. Then, we may write $X(f)$ as $X(a), a = (a_1, a_2, \dots)$.

Again, we take a Lévy's Brownian motion $X(a), a \in l^2$, which is a slight generalization of what we have discussed before. It is defined by the conditions in the same expressions as in Definition 1.1. The $X(a)$ satisfies many properties as generalizations of that in Definition 1.1. We now pause to state one of the crucial difference from the finite dimensional parameter case; namely it is the deterministic properties (see Lévy [49, 50]).

Theorem 4.1 *Let o be fixed arbitrarily. Let $X(a)$ be given in the region $V = \{a \in l^2; r_1 \le \rho(a, o) \le r_2\}$. Then, $X(a)$ is determined for every $a \in l^2$ such that $\rho(a, o) < r_1$.*

For the proof, we refer to P. Lévy [49].

Related to this surprising fact, it is interesting to observe that deterministic property of a Brownian motion increases as the dimension of the parameter space increases. For instance, we take the $M_{2p+1}(t)$ process which is the average of $(2p + 1)$-dimensional parameter Brownian motion over a sphere with radius t. As is known, the $M_{2p+1}(t)$ process is historically interesting example of a Gaussian process which has canonical representation. It is also known that multiple Markov property or analytic property increases as p increases. So, the above theorem is not surprising.

We recall that the $M_{2p+1}(t)$ is differentiable in t as many times as p, and the derivatives have some connection with the variation of $X(a)$. We can see the deterministic property increases as d increases. As for the $M_{2p}(t)$ we can discuss in connection with the innovation.

4.3 White noise parameterized by a point of a manifold

Let $W(u), u \in R^d$, be a white noise. Its probability distribution is denoted by $(E^*, \mathbf{B}, \mu^d)$, where E^* is the dual space of a sigma-Hilbert nuclear space E, which is a dense subspace of $L^2(R^d)$, and where $\mathbf{B}$ is a topological Borel field which is generated by all open subset of E^*.

Remind that the characteristic functional $C(\xi), \xi \in E$, of W is of the form

$$C(\xi) = \exp\left[-\frac{1}{2}\|\xi\|^2\right], \quad \xi \in E.$$

Let C be an ovaloid and let (C) be a domain enclosed by C. The white noise with the parameter restricted to (C) can be defined as follows:

Let $E(C)$ be a nuclear space of C^∞-functions with supports in (C). Then, the white noise $W^C(u), u \in (C)$ can be defined simply as the restriction of the parameter u to (C). The probability distribution of W^C is determined by the characteristic functional $C^C(\xi)$ which is of the form

$$C^C(\xi) = \exp\left[-\frac{1}{2}\int_{(C)} \xi(u)^2 d\sigma(u)\right], \tag{4.3.1}$$

where $d\sigma$ is the volume element on (C).

The $C^C(\xi)$ is a characteristic functional derived from $C(\xi)$ and it defines uniquely a probability measure μ^C on $E(C)^*$, which is viewed as a marginal distribution of μ^d.

Another marginal distribution is introduced by taking the basic nuclear space to be the set of all C^∞-functions on the closure K of the domain (C). The differentiability on the boundary is understood as that from the inward direction. The probability distribution μ^K of white noise $W^K(u), u \in K$, can now be introduced. The characteristic functional of μ^K is given by

$$C^K(\xi) = \exp\left[-\frac{1}{2}\int_K \xi(u)^2 du\right]. \tag{4.3.2}$$

As soon as we come to the ovaloid C, the discussion becomes somewhat complicated. An easy part is as follows. Since $W^d(u)$ has independent value at every point u, so are the systems $\{W^d(u), u \in (C)\}$ and $\{W^d(u), u \in C = K - (C)\}$, the proof of which comes easily from the observation of the characteristic functional.

4.4 Random fields parameterized by a manifold C

The study of a random field $X(C)$ with parameter C is our main purpose, and we are now ready to come to this topic. Our main results will be illustrated in and after Chapter 5.

We now list the topics to be discussed hereafter, so that a quick overview can be given.

(1) Background of the classical (non-random) variational calculus will be touched upon briefly in Appendix, Section A.2.

(2) Some significant classes of random fields. By a *random field* we mean a system of random variables on a probability space which are

parameterized by a member in one of the following spaces:

(i) A higher dimensional topological space, like R^d, $S^d, d > 1$, or subdomain of them.
(ii) An infinite dimensional topological vector space, like $L^2(R)$, l^2, etc.
(iii) A set $\mathbf{C}$ of smooth ovaloids C in a Euclidean space.

The tools from analysis depend on the class of random fields, but the probabilistic idea in this monograph is basically the same; namely the *innovation approach.*

Classes in which we are interested are in order.

(a) Gaussian random fields including particular fields, which are Gaussian processes. These will be discussed in Chapter 5.

(b) General random fields which are functionals of white noise. Stochastic processes and random fields of the form $X(C)$ are involved. We apply the S-transform to get ordinary functionals of $\xi \in E$ depending on t or on C, so that classical functional analysis can be applied.

(c) Stochastic variational equations for $X(C)$. Formulation of the equation and the existence of solution are discussed.

(d) The variation $\delta X(C)$ of $X(C)$ for infinitesimal deformations of C is discussed. Special type of deformations of C is emphasized, namely conformal transformations of the parameter space. Lie group properties of those transformations are useful. Applications to quantum dynamics are also significant.

(e) It is not too much to say that random fields as functionals of Poisson noise are as important as those of (Gaussian) white noise. Sometimes fields of Poisson noise can be dealt with in a similar manner to the Gaussian case, but dissimilarity is also interesting. We shall often discuss in parallel with these two cases and sometimes separately. Linear processes (fields) which are mixtures of the two noises, Gaussian and of Poisson type, present interesting probabilistic meaning.

(f) Random field parameterized by a closed string
Consider a random field $X(C)$ parameterized by a string C (in space). We assume it is a generalized functional of white noise. Let $U(C, \xi)$ be the S-transform of $X(C) = X(C, x)$, where x stands for a white noise. As usual C is assumed to be a smooth ovaloid. An analytic expression of C is denoted by

$$C = \vec{r}(s); \ \ \vec{r}(s) = (u(s), v(s), w(s)), \ \ 0 \le s \le S,$$

where s denotes the arc length and the string is not necessarily closed. Thus, $X(C)$ has a representation of the form

$$X(\vec{r}(s)) = X(\vec{r}(s), x), \quad x : \text{white noise.}$$

Accordingly, S-transform can be expressed in the form

$$U(\vec{r}(s)) = U(u(s), v(s), w(s)) = U(\vec{r}(s), \xi),$$

and it is ready to be analyzed.

This is a special case of (b) and particularly important in the application to quantum dynamics. However, the study of such fields has to wait for another report.

(g) Conditional expectations for the study of the Lévy Brownian motion $X(a), a \in R^d$.

Let $X(C)$ be a random field parameterized by the smooth convex contour C in the plane, and assume that it comes from the Lévy Brownian motion.

A good example of such an $X(C)$ is given by the conditional expectation of a two-dimensional parameter Lévy's Brownian motion $B(a), a \in R^2$. More precisely, let C be a convex C^∞-contour, and let p be a fixed point inside of C. Define $X(C)$ by the conditional expectation

$$X(C) = X(C, p) = E(X(p)|X(a), a \in C). \tag{4.4.1}$$

(See Chapter 1.) Concerning this $X(C)$, a generalization of the results by Si Si [70], 1987, can be given. Also, variational calculus for $X(C)$ can be discussed. While we are doing so, some profound analytic properties of $B(a)$ have been discovered.

In general, we discuss conditional expectations like (4.4.1) for general surfaces or even part of them. There it is noted that if C has boundary ∂C, then, the expression of the variation of the conditional expectation exhibits singularities of the kernel function in the integral of $B(a)$.

It may be thought that the normal derivatives along the surface C also give a system of random variables, that is a random field. But this may not be true even in the case of the Lévy Brownian motion, actually depends on d.

It is anyhow fruitful for the study of a Brownian motion $X(a)$ to construct various fields and to consider their variations.

(h) Restriction of parameters. This problem will be discussed whenever it is necessary. There are a few methods to restrict the parameter to a subdomain or a lower dimensional space. Restriction to a lower dimensional ovaloid is indispensable when exact expression of the variational formula is required. This problem will be solved according to circumstances.

4.5 Random fields as white noise functionals

In what follows, we have an overview on random fields $X(C)$ that are always assumed to be functionals (sometimes, generalized functionals) of a white noise $x(u)$. Namely,

$$\{X(C) = X(C,x);\ \ C \in \mathbf{C}\}$$

where

$$\mathbf{C} = \{C; \text{convex, diffeomorphic to } S^{d-1}\}. \tag{4.5.1}$$

Note. For the definition of $\mathbf{C}$, convexity of C is required. The reason for this requirement will be illustrated when the variation of $X(C)$ is computed, where the convexity assumption for C guarantees the possibility to define a white noise with parameter set C.

We further assume that

$$X(C) \in (L^2).$$

Hence, the S-transform can be applied to $X(C)$. Let the S-transform be denoted by $U(C,\xi)$:

$$U(C,\xi) = S(X(C))(\xi). \tag{4.5.2}$$

We are therefore ready to appeal to the classical theory of functional analysis. If necessary, C may be represented by a vector valued function.

(1) Linear functional of x

In this case, the $X(C)$ is assumed to be expressed in the form

$$X(C) = \int_{R^d} F(C,u)x(u)du^d.$$

An interesting class involves such integrals as

$$X(C) = \int_{(C)} F(C,u)x(u)du^d,$$

where (C) is the domain enclosed by C.

The expression above is called a *causal* representation of $X(C)$ in terms of white noise. The term causal comes from the part that, if C is thought of as present, then $x(u), u$ being inside of C, are past values.

Topics on such random fields $X(C)$ will be discussed in the next chapter. Various interesting properties are found by analogy with the canonical representation of a Gaussian process $X(t)$.

(2) Nonlinear functionals living in (L^2)

By using the Fock space decomposition of (L^2), we have, in general, the decomposition of $X(C) = X(C, x)$:

$$X(C) = \sum_{0}^{\infty} X_n(C),$$

where $X_n(C) \in H_n$, and $X_n(C)$ is a homogeneous chaos of degree n. We are therefore suggested to take a homogeneous chaos parameterized by C separately.

Another remark is that a field $X(C)$ extends to a generalized white noise functional.

Throughout these cases, the idea of the innovation approach is sitting behind.

4.6 Random fields as Poisson noise functionals

Poisson noise with R^d-parameter has been defined, and random fields as functionals of Poisson noise are dealt with. If ordinary fields have finite variance, then the fields in question can be discussed in Hilbert space $(L^2)_P$ like Gaussian case. However we often meet the case where we have to analyze without assuming the existence of variance. In such a case a wider space $(\mathbf{P})$ has been introduced in Section 3.9 for our purpose. Characteristic functional for a stochastic process is another tool from analysis, and operations like subordination and jump finding of sample functions can be discussed. Further, random fields expressed as functionals, either linear or nonlinear, of Poisson noise are well investigated in this line.

Chapter 5

Gaussian Random Fields

This chapter is devoted to the study of the Gaussian random fields, where detailed and profound results are stated with special emphasis on their innovations. We start with the case of a Gaussian process, that is, the case depending on of one-dimensional parameter t. The canonical representation theory of a Gaussian process has been established and rather well-known (see [17]), however we should note that the significance of the representation theory has been rediscovered not only in stochastic process theory, but also in quantum dynamics, molecular biology and other fields of applications.

One of the main topics in this chapter is the Markov property of a random field, which is the most significant notion to express the dependence on the parameter. The multiple Markov property of a Gaussian process has been introduced many years ago (see [17]), and the significance of this notion has now been recognized and one meets the revival of the study of the random complex system that enjoy the property.

Our interest is in a generalization of the canonical representation theory for $X(t)$ to a certain class of Gaussian random fields $X(C)$ that are causally represented in terms of multi-parameter white noise. This theory serves as a background of the study of the Markov property.

5.1 A review of the canonical representations of Gaussian processes

Let $X(t), t \geq 0$, be a Gaussian process with $E[X(t)] = 0$. If there is a white noise $\dot{B}(t)$ such that $X(t)$ is expressed in the form

$$X(t) = \int_0^t F(t,u)\dot{B}(u)du, \tag{5.1.1}$$

with a kernel $F(t,u)$ of Volterra type, then the expression is called a *representation* of $X(t)$, more precisely, a *causal* representation in terms of white noise. A representation is determined by the pair $\{F(t,u), \dot{B}(u)\}$, so that the pair itself is often called the representation of $X(t)$.

To be surprised, the representation is not unique. Before an exact statement on this assertion is given, it is necessary to give a definition of the equivalence of the representations.

Definition 5.1 Two representations $\{F_i(t,u), \dot{B}_i(u)\}, i = 1, 2$, are called equivalent if and only if for $u \leq t$

$$F_1(t,u)^2 = F_2(t,u)^2 \quad \text{a.e. for every } t. \tag{5.1.2}$$

It is easy to see that there are many nonequivalent representations for every Gaussian process (see [17]). Among others there is a particular one called the canonical representation.

Definition 5.2 A representation $\{F(t,u), \dot{B}(u)\}$ is called *canonical* and $F(t,u)$ is called a canonical kernel if the conditional expectation satisfies the following condition holds:

$$E[X(t)|\mathbf{B}_s(X)] = \int_0^s F(t,u)\dot{B}(u)du, \quad s \leq t. \tag{5.1.3}$$

Theorem 5.1 *The canonical representation is unique up to equivalence if it exists.*

For the proof of this theorem and a kernel criterion for the canonical representation (canonical kernel) we refer to the literature [17].

The significance of the canonical representation is that various probabilistic properties of a Gaussian process can be expressed by the known properties of white noise (or Brownian motion) and by the analytic properties of the kernel function. As for the kernel the stationarity, the multiple Markov property, the reversibility and others can be expressed in terms of the function properties of the uniquely determined canonical kernel.

Definition and many of the properties of the representation of Gaussian random fields are inherited from those of a Gaussian processes $X(t)$. We shall see this situation in the next section.

5.2 Canonical representations of Gaussian random fields

Let $X(C)$ be a Gaussian random field parameterized by C. In particular, $X(C)$ is assumed to be a linear function of the white noise x. The class $\mathbf{C}$

of the family of C is taken to be as follows:

$$\mathbf{C} = \{C; C : \text{closed convex surface in } R^d, \text{diffeomorphic to } S^{d-1}\}. \quad (5.2.1)$$

The class $\mathbf{C}$ is topologized by the Euclidean metric. We sometimes take a sub-class of $\mathbf{C}$, depending on the subject to be discussed.

A smooth, closed, convex surface is often called ovaloid, which will frequently be used in what follows.

Assume that

(i) $X(C) \neq 0$ a.e. for every C, and $E[X(C)] = 0$.

(ii) Let $\Gamma(C, C') = E[X(C)X(C')]$ be the covariance function of $X(C)$. Then, $\Gamma(C, C')$, for $C > C'$, admits to take variation in the variable C and $\Gamma(C, C')$ never vanishes.

Our attention will be focussed on the Gaussian random field $\{X(C); C \in \mathbf{C}\}$, with a *causal* representation of the form

$$X(C) = \int_{(C)} F(C, u)x(u)du, \quad (5.2.2)$$

in terms of the R^d-parameter white noise $x(u)$ and of $L^2(R^d)$-kernel $F(C, u)$ for every C. As before (C) denotes the domain enclosed by C.

A remark should be given to the term "causal". Given a surface C, then its inside is therefore understood to express the *past*, so that the outside is the *future* and C itself is tacitly understood to represent the *present.* Hence, the causal representation means that the given field is a function only of the past values of x. This is a generalization of the causality in the case of the time parameter only.

The notion of a canonical representation of a Gaussian random field $X(C)$ can be defined in a similar manner to the case of Gaussian process

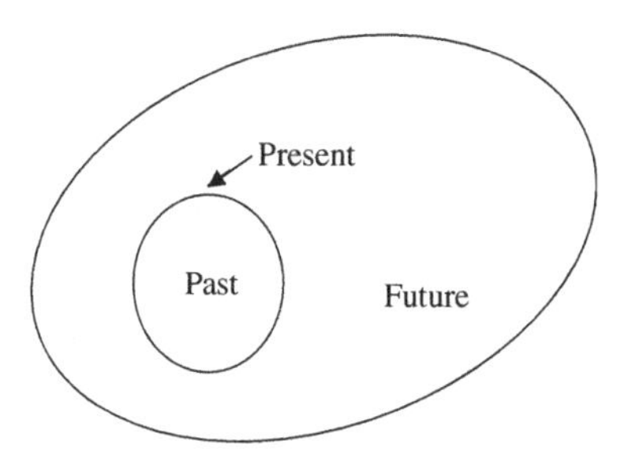

Fig. 6. For causality.

$X(t)$ (see [17]). Accordingly, similar results are obtained, as is expected. In many of the important cases we are given those fields that have causal representation in terms of white noise.

Definition 5.3 Let $\mathbf{B}_{C'}(X)$ be the smallest sigma field generated by $\{X(C), C < C'\}$. The notation $C < C'$ stands for the relationship that the domain (C) enclosed by C is included in C'. The representation (5.2.2) is called a *canonical representation* if

$$E[X(C)|\mathbf{B}_{C'}(X)] = \int_{(C')} F(C,u)x(u)du, \tag{5.2.3}$$

holds for any pair (C', C) with $C' < C$.

Remark 5.1 *Sometimes the class of* C*'s is taken to be a subclass of* $\mathbf{C}$. *Even in such a case we can speak of canonical representation. If nothing is mentioned,* $\mathbf{C}$ *is always assumed to contain all possible* C*'s.*

The equivalence of the representation can be defined in the similar manner to the case of a process. Up to the equivalence we claim

Theorem 5.2 *The canonical representation is unique if it exists.*

Proof. Take the variance of the conditional expectation, given in (5.2.3).

$$E\left[E[X(C)|\mathbf{B}_{C'}(X)]^2\right] = \int_{(C')} F(C,u)^2 du, \quad C' < C. \tag{5.2.4}$$

We should note that the variance depends only on the probability distribution of $\{X(C)\}$ and is independent of the choice of representation.

More precise observation is as follows. If the representation is not unique, there are two canonical kernels F and F^* and then

$$\int_{(C')} F(C,u)^2 du = \int_{(C')} F^*(C,u)^2 du$$

holds for any $C' < C$. Hence we have

$$F(C,u) = \varepsilon(C,u)F^*(C,u); \quad |\varepsilon(C,u)| = 1, \tag{5.2.5}$$

where ε is a measurable function of u for every C.

By the two kernels F and F^*, the covariance of (5.2.3) is obtained as

$$E\big[E[X(C)|\mathcal{B}_{C''}(X)]E[X(C')|\mathcal{B}_{C''}(X)]\big] = \int_{(C'')} F(C,u)F(C',u)du,$$

where $C'' < C$ and $C'' < C'$.

On the other hand, we obtain

$$\begin{aligned} &E\{E[X(C)|\mathcal{B}_{C''}(X)]E[X(C')|\mathcal{B}_{C''}(X)]\} \\ &= \int_{(C'')} F^*(C,u)F^*(C',u)du \\ &= \int_{(C'')} F(C,u)F(C',u)\varepsilon(C,u)\varepsilon(C',u)du \end{aligned}$$

by (5.2.5).

Similarly for any $C''' \in \mathbf{C}$; $C''' < C'$, we have

$$\int_{(C''')} F(C,u)F(C',u)du = \int_{(C''')} F(C,u)F(C',u)\varepsilon(C,u)\varepsilon(C',u)du.$$

Thus the equality

$$F(C,u)F(C',u) = F(C,u)F(C',u)\varepsilon(C,u)\varepsilon(C',u)$$

holds almost everywhere. We can see that, for u such that $F(C,u)\ F(C',u) \neq 0$,

$$\varepsilon(C,u)\varepsilon(C',u) = 1, \text{ on } (C').$$

Fix $C = C_0$, and determine $\varepsilon(C_0,u)(=\pm 1)$ as a function of u. Thus

$$\varepsilon(C',u) = \frac{1}{\varepsilon(C_0,u)} = \varepsilon(C_0,u), \quad \forall C'.$$

This means that $\varepsilon(C,u)$ is independent of C.

Thus it is proved that $F(C,u)$ is unique up to ± 1.

The existence of canonical representation

We assume that a given Gaussian random field $X(C), C \in \mathbf{C}$, is centered and is separable. Let $M_t(X)$ be the subspace of $L^2(\Omega, P)$ spanned by all the $X(C)$ with $C < C_t$, where C_t is a sphere with radius t. For convenience, we make $M_t(X)$ to be right continuous, in fact we consider $M_{t+}(X) = \lim_{\epsilon \to 0+} M_{t+\epsilon}(X)$, which will be denoted by $M_t(X)$. Obviously $M_t(X)$ is right continuous in t. Our essential assumption is that the $X(C)$ has no remote past:

$$\bigcap M_t(X) = \{0\}.$$

Associated with $M_t(X)$ is a projection $E(t)$; indeed the family $\{E(t)\}$ is a resolution of the identity I in $\bigvee M_t(X) = M(X)$.

Now we can appeal to the Hellinger–Hahn theorem. As a result, we are given atmost countably many white noises $W_n(t)$. Take a system of spherical harmonics $\{Y_k(\theta)\}$ and define a family of white noises parameterized by θ for every t. Set, for each t,

$$W(t,\theta) = \sum_0^\infty Y_k(\theta) W_k(t).$$

A white noise is given by a consistent family $\{W(u), u = (t,\theta)\}$. Since $\{W(u), t \le s\}$ has the same information as that of $M_t(X)$, and it is easily seen that $X(C)$ is expressible as an integral of the form

$$X(C) = \int_{(C)} F(C,u) W(u) du^d,$$

with a suitable kernel $F(C,u)$. Details will be discussed in the forthcoming paper by the authors.

Theorem 5.3 *Under the assumptions that $E(X(C)) = 0$, separable and no remote past, the $X(C)$ with a causal representation has the canonical representation.*

Kernel criterion for canonical representation of Gaussian random field

Following the kernel criterion for canonical representation of Gaussian processes, we can give that of Gaussian random field.

Assume that

(i) *$X(C)$ has a causal representation*
(ii) *there is no open set G such that $\int_G F(C,u)\varphi(u)du = 0$ for any φ with* $\mathrm{supp}\{\varphi\} \subset G$.

Theorem 5.4 *A random field $X(C)$, satisfying the above assumptions, has canonical representation if and only if for all $(C) \subset (C_1)$; C_1 being fixed,*

$$\int_{(C)} F(C,u)\varphi(u)du = 0 \quad \textit{implies} \quad \varphi(u) = 0 \quad \textit{a.e. on } (C_1)$$

holds.

Proof. Since we are concerned with Gaussian system, $E[X(C)|\mathcal{B}_{C'}]$ is the projection of $X(C)$ down to the closed linear space spanned by $\{X(C'');$ $C'' < C'\}$.

By assumption the closed linear space $M_{C'}(X)$, spanned by $\{X(C''), C'' < C'\}$ is the same as the closed linear subspace $M_{C'}(x)$ spanned by $\{x(u); u \in (C')\}$.

The reason is that $M_{C_0}(X) \subset M_{C_0}(x)$, in general, since $X(C'')$ is a (linear) function of $x(u); u \in (C'')$.

If $M_{C_0}(X) \neq M_{C_0}(x)$ then there exist $\varphi \neq 0$ a.e. on a set of positive measure such that $\int_{C_0} \varphi(u)x(u)du$ is orthogonal to $X(C''); C'' < C_0$. It contradicts to the assumption. Thus the assertion is proved.

Here we illustrate the canonical and noncanonical properties of a representation in the following examples.

Example 5.1 $X(C) = \int_{(C)} x(u)du, C \in \mathbf{C}$, where $\mathbf{C}$ is given by (5.2.1), is a canonical representation. In fact, it is a martingale (see Section 5.3).

Example 5.2 Consider a random field $X(C); C \in \mathbf{C_0}$, where $\mathbf{C_0}$ is a family of circles, with the representation

$$X(C) = X_0 \int_{(C)} e^{-k\rho(C,u)} x(u)\nu(u)du,$$

where ρ denotes the distance, k is a constant and ν is a given continuous function. We can easily prove that it is a canonical representation.

Indeed, it is the solution of Langevin equation,

$$\delta X(C) = -X(C) \int_C k\delta n(s)ds + X_0 \int_C \nu(s)x(s)\delta n(s)ds,$$

where $C \in \mathbf{C}_0$.

Example 5.3 Let $\{C_R, R \in \mathbf{R}\}$ be a family of concentric circles C_R with radius R and with center at the origin. Then

$$X(C) = \int_{(C_R)} (3R - 4|u|)x(u)du$$

is a noncanonical representation of $X(C)$, since there is a function $\varphi(u) = |u| \neq 0$ such that

$$\int_{(C_R)} (3R - 4|u|)\varphi(|u|)du = 0.$$

Even in the non-canonical case, we always have the same absolute value of the kernel on the boundary C, namely the value $|F(C,s)|, s \in C$, is the same, almost everywhere on C for any representation.

5.3 Martingale

Under the understanding of past, present and future that are explained in the last section, we can define the notion of a martingale.

Definition 5.4 Let $\mathbf{B}_C(x) = \mathbf{B}\{\langle x, \xi \rangle; \text{supp}\{\xi\} \subset (C)\}$. If

(1) $E|X(C)| < \infty$ and

(2) $E[X(C)|\mathbf{B}_{C'}(x)] = X(C')$, for any $C' < C$,

then $X(C)$ is called a martingale with respect to $\mathbf{B}_C(x)$.

Theorem 5.5 *If a Gaussian random field $Y(C)$, with mean zero, has a canonical representation and is a martingale, then there exists a locally square integrable function g such that*

$$Y(C) = \int_{(C)} g(u)x(u)du. \tag{5.3.1}$$

Proof. Since $Y(C)$ has a canonical representation in terms of a white noise x, it is expressed in the form

$$Y(C) = \int_{(C)} g(C, u)x(u)du. \tag{5.3.2}$$

Take the conditional expectation $E[Y(C)|\mathbf{B}_{C'}(x)]$, $C' < C$. It is equal to $Y(C')$ by the martingale property. On the other hand, the above representation (5.3.2) is canonical, so that we have

$$Y(C') = \int_{(C')} g(C, u)x(u)du.$$

Thus we have

$$\int_{(C')} g(C', u)x(u)du = \int_{(C')} g(C, u)x(u)du.$$

It follows that

$$g(C', u) = g(C, u)$$

almost everywhere on (C'). That is, if $(C) \supset (C')$ then $g(C, u) = g(C', u)$ on (C'). Thus, the inductive limit of $g(C, u)$ can be written as $g(u)$ and the restriction of g on (C) is

$$g(u)|_{(C)} = g(C, u).$$

The assertion is proved.

Proposition 5.1 *If $Y(C)$ is a martingale, never vanishes for every C and if $Y(C, x)$ is in the space (S), then $\mathbf{B}_C(x) = \mathbf{B}_C(Y)$.*

Proof. Since $Y(C)$ is a martingale, it can be expressed as

$$Y(C) = \int_{(C)} g(u)x(u)du. \tag{5.3.3}$$

Hence

$$E[Y(C)^2] = \int_{(C)} g(u)^2 du \neq 0, \quad \textit{for every } C.$$

It means that $g \neq 0$ almost everywhere.

Take the variation (refer to Chapter 7) of $Y(C)$:

$$\delta Y(C) = \int_C g(s)\delta n(s)x(s)ds. \tag{5.3.4}$$

Let δn vary, then we have $\{g(s)\delta n(s)\}$ which is rich enough to determine $\{x(s)\}$. Thus $x(s)$ can be obtained from $\{Y(C'), C' \leq C\}$. It follows that $\mathbf{B}_C(x) \subset \mathbf{B}_C(Y)$. The converse relation is obvious and the assertion is proved.

5.4 A review of Markov property of Gaussian processes

We are now in a position to define multiple Markov property of Gaussian random fields $X(C)$. Before doing so, it is necessary to remind the definition of *multiple* Markov property of a stochastic process $X(t)$, in fact, to remind how we have come to this way of defining the properties. We now pause to explain the story behind the definition.

Some thought on Markov property behind the definition

The simple Markov property is quite well known. A notion of multiple Markov property should also express the dependence of the random variables $X(t)$'s as t runs, in a more general sense than simple Markov property. Such a generalized notion is naturally to be introduced in connection with the prediction theory, namely related to the property that shows how the past and the future are related in a complex manner.

Let t denote the present instant. One may consider how much information obtained by the past values of $X(s), s \leq t$, is necessary to get the best predictor of the future value $X(t+h), h > 0$.

Theoretically, one can use all the information contained in the observed data to predict a future value. In a favourable case, finitely many, say as many as N, random variables formed by the observed values in the past are necessary and sufficient to get the best prediction of a future value.

To make the situation homogeneous in time, the number N is determined uniformly in t. How large the N can be is one of the characteristic of the dependency of the process in question.

The property that the number N is finite will reflect upon the computability of the prediction.

Let N be fixed. Consider a particular case where we need $X(t_j)$'s as many as N, with t_j's infinitesimally close to t from the left. In such a case we may assume that $X(t)$ is differentiable (in the $L^2(\Omega)$-topology or sample function-wise) up to $N-1$ times. The function $\varphi(X(t), X'(t), \ldots, X^{(N-1)}(t), t, h)$ of these members can give the best prediction of a future value $X(t+h)$ of time $h > 0$ ahead. The random elements to determine $X(t+h)$ would be $X^{(j)}(t)$, $j = 0, 1, \ldots, N-1$, and innovation.

Now we must note an essential remark. In principle, to define the dependence, it is not reasonable to assume the differentiability of $X(t)$. It is therefore quite natural to come to the property that a necessary and sufficient information for the prediction is obtained by finite number of $\mathcal{B}_t(X)$-measurable random variables, not necessarily $X^{(j)}(t)$'s. For further interpretation we need mathematical formulas.

If we focus our attention on Gaussian processes, everything related to Markov property is consistently formulated with mathematical rigor. The *multiple Markov Gaussian process* has been defined in [17] 1960, and it satisfies all the requirements that have been requested so far. It is as follows.

Let $X(t), t \in T$, T being an interval, be a separable Gaussian process with $E[X(t)] = 0$.

Definition 5.5 Let $X(t), t \in [0, \infty)$, with $E[X(t)] = 0$ be a Gaussian process. If, for any $t_0 \le t_1 < t_2 < \cdots < t_N < t_{N+1}$,

(1) $E[X(t_i)|\mathcal{B}_{t_0}(X)], i = 1, 2, \ldots, N$ are linearly independent and if

(2) $E[X(t_i)|\mathcal{B}_{t_0}(X)], i = 1, 2, \ldots, N+1$ are linearly dependent,

then $X(t)$ is called N-ple Markov Gaussian process.

Theorem 5.6 *If $X(t)$ is N-ple Markov and if its canonical representation is given by*

$$X(t) = \int_0^t F(t,u)\dot{B}(u)du, \tag{5.4.1}$$

then $F(t,u)$ is a Goursat kernel of the form

$$F(t,u) = \sum_{i=1}^{N} f_i(t)g_i(u),$$

where $\det(f_i(t_j)) \neq 0$, for different t_j's, and where g_i's are linearly independent in $L^2([0,t])$ for any $t > 0$.

If in addition, $X(t)$ is assumed to be a stationary process with $E[X(t)] = 0$, then the Goursat kernel is a function of $t - u$, with $u \leq t$. More concretely, we have

Theorem 5.7 *If $X(t)$ is N-ple Markov and mean continuous stationary Gaussian process with no remote past, then it has the canonical representation with kernel $F(t,u) = F(t-u)$ and the Fourier transform $\hat{F}(\lambda)$ of $F(u)$ is of the form*

$$\hat{F}(\lambda) = \frac{Q(i\lambda)}{P(i\lambda)},$$

where P and Q are polynomial and the degree of P is N and that of Q is less than N.

Note. Proofs of Theorem 5.6 and 5.7 are referred to [17].

As another particular case, $X(t)$ is assumed to be $N-1$ times differentiable, and satisfies

$$\sum_{k=0}^{N} a_k(t)X^{(k)}(t) = \dot{B}(t),$$

where $X^{(N)}(t)$ is understood as a generalized process. More precisely, define a differential operator

$$L_t = \sum_{k=0}^{N} a_k(t)\frac{d^k}{dt^k}$$

and let L_t^* be the formal adjoint operator of L_t. If the bilinear form $\langle L_t^*\xi, X\rangle$ is well defined and if its characteristic functional of $L_tX(t)$ is $\exp[-\frac{1}{2}\|\xi\|^2]$, then we may write as above.

Then, $X(t)$ has the canonical representation of the form

$$X(t) = \int_0^t R(t,u)\dot{B}(t)dt,$$

where $R(t,u)$ is the Riemann function for the differential operator L_t.

Note that $X^{(N)}(t)$ is understood to be the right derivative of $X^{(N-1)}(t)$, so that it is not $\mathcal{B}_{t+}(X)$-measurable, on the contrary so are $X^{(k)}(t)$, $0 \le k \le N-1$.

Such a process $X(t)$ is known as an *N-ple Markov Gaussian process in the restricted sense. It is easy to see that the $X(t)$ is N-ple Markov.*

Several remarks concerning the Gaussian Markov property are now in order.

Remark 5.2 *The multiple (N-ple) Markov property is often misunderstood because of the confusion of necessary condition with sufficient condition. Namely, whenever we take the projection of any future value of $X(t+h), h \ge 0$ to the past which is the closed linear manifold spanned by $X(s), s \le t$, it is necessary to be N-dimensional, but it is not a sufficient condition.*

Remark 5.3 *The canonical kernel may be understood, roughly speaking, to be a kernel associated to a causally invertible linear operator. Formally, the inverse operator of the integral operator F of Volterra type exists, let it be denoted by $G(= F^{-1})$ which is also of Volterra type, then, we have*

$$(GX)(t) = \dot{B}(t).$$

Hence, $\dot{B}(t)$ obtained above is nothing but the innovation of $X(t)$.

Remark 5.4 *Related to the dependence, there is a notion of "multiplicity" of a Gaussian process $X(t)$. Suppose the multiplicity of $X(t)$ is M. Then there are M independent additive Gaussian processes, the increments of them over the same time interval are independent. It is surprising, but in a sense natural. An example with higher multiplicity has been given by M. Hitsuda (see [38]). It is noted that, like multiple Markov property, the analytic property of the $X(t)$ does not contribute to have higher multiplicity, but it is used to see the complexity of the given Gaussian processes.*

If one observes the expressions appeared so far, it is natural to consider a generalization to non-Gaussian processes. Interesting examples are

1. $\dot{B}(t)dt$ is replaced by an orthonormal random measure $dZ(t)$, and $X(t)$ is given by a pair of a non-random kernel $F(t,u)$ and $dZ(t)$:

$$X(t) = \int^t F(t,u)dZ(u).$$

   Assume that $\frac{dZ(t)}{dt}$ is a Poisson noise. Note that, in this case, there is another way of understanding the above integral, namely, $dZ(t)$ is viewed as a Stieljes measure defined by a sample function of Poisson process. Hence, we may consider multiple Markov property in another way different from Gaussian case.
2. An orthogonal random measure may be taken to be a homogeneous chaos, that is defined by Wick product $: \dot{B}(t_1) \cdots \dot{B}(t_n) :$. It is a multi-dimensional orthogonal random measure.

5.5 Markov property of Gaussian random fields

As before $X(C)$ is assumed to be expressed in the form (5.2.2).

A. Markov property

The class $\mathbf{C}$ is taken to be as before.

Definition 5.6 If a random field $X(C)$ satisfies the equality

$$P(X(C) \in B|\mathbf{B}_{C'}(X)) = P(X(C) \in B|X(C')) \tag{5.5.1}$$

for every $C \geq C'$, the $X(C)$ is called a Markov field.

In our case, $\{X(C)\}$ is Gaussian, so that it is sufficient to define the Markov property by

$$E[X(C)|\mathbf{B}_{C'}(X)] = E[X(C)|X(C')], \quad C' \leq C. \tag{5.5.2}$$

Theorem 5.8 *Assume that $X(C) \neq 0$ satisfies the Markov property, then there exists $f \neq 0$ and $Y(C)$ which is a martingale with respect to $\mathbf{B}_C(Y)$ such that*

$$X(C) = f(C)Y(C). \tag{5.5.3}$$

Proof. Let $C'' \leq C' \leq C$. Since $X(C)$ satisfies the Markov property, we have

$$E[X(C)|\mathbf{B}_{C'}(X)] = E[X(C)|X(C')].$$

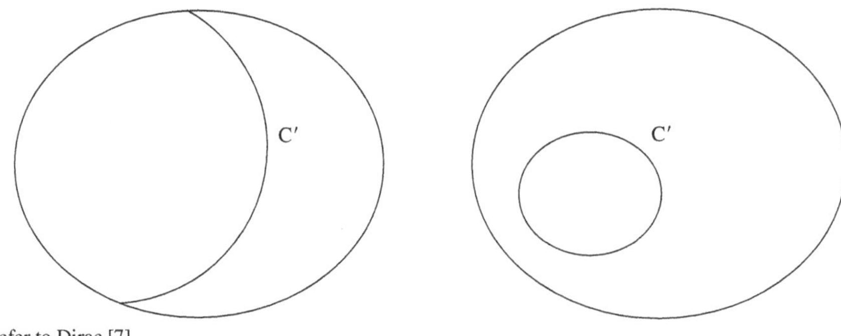

Fig. 7. Markov property.

Since $\{X(C), X(C')\}$ is Gaussian there is a non-random $\psi(C, C') \neq 0$ such that

$$E[X(C)|X(C')] = \psi(C, C')X(C'), \quad C' < C.$$

Hence

$$E[X(C)|\mathbf{B}_{C'}(X)] = \psi(C, C')X(C'), \quad C' < C. \tag{5.5.4}$$

Take the conditional expectation $E[\,\cdot\,|\mathbf{B}_{C'}(X)]$ of both sides of (5.5.4) where $C'' < C'$, then

$$E\big[E[X(C)|\mathbf{B}_{C'}(X)]|\mathbf{B}_{C''}(X)\big] = \psi(C, C')E[X(C')|\mathbf{B}_{C''}(X)].$$

Thus we have

$$E[X(C)|\mathbf{B}_{C''}(X)] = \psi(C, C')\psi(C', C'')X(C'').$$

The left hand side is equal to $\psi(C, C'')X(C'')$. Hence we have

$$\psi(C, C'') = \psi(C, C')\psi(C', C''). \tag{5.5.5}$$

For $C' \geq C$, we can define $\psi(C', C) = 1/\psi(C, C')$, since $\psi(C, C') \neq 0$. Thus, $\psi(C, C')$ is defined for every pair (C, C') with $C' < C$, and hence

$$\psi(C, C'') = \frac{\psi(C', C'')}{\psi(C', C)}. \tag{5.5.6}$$

For fixed C', define $f(C) = \psi(C, C')$. Note that this definition depends on the choice of C'. However, $f(C)$ can be determined up to constant.

Define

$$Y(C) = \frac{X(C)}{f(C)}. \tag{5.5.7}$$

It is easily seen that $Y(C)$ is martingale with respect to $\mathbf{B}_C(Y)$. Thus the assertion is proved.

Corollary 5.1 *If $Y(C,x) \in (S)$ then $Y(C)$ is a martingale with respect to $\mathbf{B}_C(X)$, as well.*

B. Multiple Markov property

Definition 5.7 For any choice of C_i's such that $C_0 \leq C_1 < \cdots < C_N < C_{N+1}$, if

i. $E[X(C_i)|\mathcal{B}_{C_0}(X)], i = 1, 2, \ldots, N$ are linearly independent and

ii. $E[X(C_i)|\mathcal{B}_{C_0}(X)], i = 1, 2, \ldots, N+1$ are linearly dependent

then $X(C)$ is called N-ple Markov Gaussian random field.

Theorem 5.9 *If $X(C)$ is an N-ple Markov and if it has a canonical representation, then it is of the form*

$$X(C) = \int_{(C)} \sum_1^N f_i(C) g_i(u) x(u) du, \tag{5.5.8}$$

where the kernel $\sum f_i(C) g_i(u)$ is a Goursat kernel, i.e. $\{f_i(C)\}, i = 1, \ldots, N$ satisfies

$$\det(f_i(C_j)) \neq 0, \text{ for any } N \text{ different } C_j, \tag{5.5.9}$$

and $\{g_i(u)\}, i = 1, \ldots, N$, on any (C), are linearly independent in L^2-space.

Proof. Let

$$X(C) = \int_{(C)} F(C,u) x(u) du$$

be a canonical representation of $X(C)$ where $F(C,u)$ is a canonical kernel.

According to the assumption that, $X(C)$ is an N-ple Markov process, we can prove that for any C_j with $C_1 < \cdots < C_N < C$, there exist coefficients $a_j(C; C_1, \ldots, C_N)$ such that $[X(C) - \sum a_j(C; C_1, \ldots, C_N) X(C_j)]$ is independent of $X(C'), C' \leq C_1$.

Thus we have

$$\int_{(C')} F(C',u)\left(F(C,u)-\sum_{j=1}^{N} a_j(C;C_1,\ldots,C_N)F(C_j,u)\right)x(u)du=0.$$

Since $F(C',u)$ is a canonical kernel, we have

$$F(C,u)=\sum_{k=1}^{N} a_k(C;C_1,\ldots,C_N)F(C_k,u). \tag{5.5.10}$$

Take N different $\{C'_j\}$ with $C'_1<\cdots<C'_N$, arbitrarily in the class $\mathbf{C}$. Using the expression of F as is in (5.5.10), we obtain

$$\begin{aligned}&\sum_{j=1}^{N} a_j(C;C'_1,\ldots,C'_N)F(C'_j,u)\\&\quad=\sum_{j,k=1}^{N} a_k(C;C_1,\ldots,C_N)a_j(C_k;C'_1,\ldots,C'_N)F(C'_j,u).\end{aligned}$$

The N-ple Markov property of $X(C)$ implies the linear dependence of $\{F(C'_j,u), j=1,\ldots,N\}$. Thus we have

$$\sum_{k=1}^{N} a_k(C;C_1,\ldots,C_N)a_j(C_k;C'_1,\ldots,C'_N)=a_j(C;C'_1,\ldots,C'_N), \tag{5.5.11}$$

for all j.

We can now prove that

$$\det(a_j(C_k;C'_1,\ldots,C'_N))\neq 0, \tag{5.5.12}$$

since $F(C_j,u)=\sum_{k=1}^{N} a_k(C_j;C'_1,\ldots,C'_N)F(C'_k,u)$, $k=1,\ldots,N$ are linearly independent functions. Then (5.5.11) becomes

$$\mathbf{a}(C,\underline{\mathcal{C}})=\mathbf{a}(C,\underline{\mathcal{C}}')B(\underline{\mathcal{C}}',\underline{\mathcal{C}}), \tag{5.5.13}$$

where

$$\mathbf{a}(C,\underline{\mathcal{C}})=(a_j(C;C_1,\ldots,C_N);\ j=1,\ldots,N)$$

and

$$B(\underline{\mathcal{C}},\underline{\mathcal{C}}')=\left[b_{jk}(C_1,\ldots,C_N;C'_1,\ldots,C'_N)\right]_{j,k=1,\ldots,N'}$$

with $\det(B(\underline{\mathcal{C}},\underline{\mathcal{C}}'))\neq 0$.

For any $C'_j \in \mathbf{C}$, $j = 1, \cdots, N$ such that $C'_j < C_j$,

$$\mathbf{a}(C, \underline{\mathcal{C}}) = \mathbf{a}(C, \underline{\mathcal{C}}')B(\underline{\mathcal{C}}', \underline{\mathcal{C}}) = \mathbf{a}(C, \underline{\mathcal{C}}'')B(\underline{\mathcal{C}}'', \underline{\mathcal{C}}')B(\underline{\mathcal{C}}', \underline{\mathcal{C}}),$$

and

$$\mathbf{a}(C, \underline{\mathcal{C}}) = \mathbf{a}(C, \underline{\mathcal{C}}'')B(\underline{\mathcal{C}}'', \underline{\mathcal{C}}).$$

Hence

$$B(\underline{\mathcal{C}}'', \underline{\mathcal{C}}')B(\underline{\mathcal{C}}', \underline{\mathcal{C}}) = B(\underline{\mathcal{C}}'', \underline{\mathcal{C}}). \tag{5.5.14}$$

Let us take fixed C_j's and define $\mathbf{f}_{\underline{\mathcal{C}}'}(C)$, $\underline{\mathcal{C}}' = (C'_1, \ldots, C'_N)$, by

$$\mathbf{f}_{\underline{\mathcal{C}}'}(C) = \mathbf{a}(C, \underline{\mathcal{C}}')B(\underline{\mathcal{C}}', \underline{\mathcal{C}}), \quad \text{for } C > C'_N$$

where $\underline{\mathcal{C}}'$ is any N-ple $(C'_1, \ldots, C'_N)$ such that $C_N > C_{N-1} > \cdots > C_1 > C'_N > C'_{N-1} > \cdots > C'_1$, thus we can see that $\mathbf{f}_{\underline{\mathcal{C}}''}$ is an extension of $\mathbf{f}_{\underline{\mathcal{C}}'}(C)$ if $C'_N > C'_{N-1} > \cdots > C'_1 > C''_N > C''_{N-1} > \cdots > C''_1$.

It shows that there exists a common extension denoted by $\mathbf{f}(C) = (f_1(C), \ldots, f_N(C))$ for all $\mathbf{f}_{\underline{\mathcal{C}}'}(C)$'s.

We can see, from (5.5.12) and the definition of $\mathbf{f}_{\underline{\mathcal{C}}'}(C)$, that $f_i(C)$ satisfies (5.5.9).

Fix $C_0 \in \mathbf{C}$. If $C > C_N > \cdots > C_1 > C'_N > \cdots > C'_1 > C_0$, then we have

$$\begin{aligned}
F(C, u) &= \sum_{j=1}^{N} a_j(C; C_1, \ldots, C_N)F(C_j, u) \\
&= \mathbf{a}(C, \underline{\mathcal{C}})\mathbf{F}(\underline{\mathcal{C}}, u)^* \\
&= \mathbf{f}(C)B(\underline{\mathcal{C}}', \underline{\mathcal{C}})^{-1}\mathbf{F}(\underline{\mathcal{C}}, u)^* \\
&= \mathbf{f}(C)\mathbf{g}(u, \underline{\mathcal{C}}', \underline{\mathcal{C}})^*,
\end{aligned}$$

where

$$\mathbf{F}(\underline{\mathcal{C}}, u) = (F(\mathcal{C}_1, u), \ldots, F(\mathcal{C}_N, u)),$$

$\mathbf{F}(\underline{\mathcal{C}}, u)^*$ denotes the transpose of $\mathbf{F}(\underline{\mathcal{C}}, u)$, and

$$\mathbf{g}(u, \underline{\mathcal{C}}', \underline{\mathcal{C}}) = \mathbf{F}(\underline{\mathcal{C}}, u)(B(\underline{\mathcal{C}}', \underline{\mathcal{C}})^{-1})^*.$$

For $C > C''_N > \cdots > C''_1 > C'''_N > \cdots > C'''_1$, we can write

$$F(\underline{C}, u) = \mathbf{f}(C)\mathbf{g}(u, \underline{\mathcal{C}}''', \underline{\mathcal{C}}'')^*,$$

so that

$$\mathbf{f}(C)\mathbf{g}(u,\underline{C}',\underline{C})^* = \mathbf{f}(C)\mathbf{g}(u,\underline{C}''',\underline{C}'')^*,$$

for $C > C''_N$. Since $\mathbf{f}$ satisfies (5.5.9), we have

$$\mathbf{g}(u,\underline{C}',\underline{C})^* = \mathbf{g}(u,\underline{C}''',\underline{C}'')^*.$$

Thus $g(u) = \mathbf{g}(u,\underline{C}',\underline{C})$ is well defined as a function of u, and

$$F(C,u) = \mathbf{f}(C)\mathbf{g}(u)^* = \sum_{i=1}^{N} f_i(C)g_i(u),$$

where $\{g_i(u), i = 1,\ldots,N\}$ are linearly independent since $\{F(C_j,u)\}$; $j = 1,\ldots,N$, are linearly independent. Thus, the theorem is proved.

Corollary 5.2 *If $X(C)$ is an N-ple Markov Gaussian random field, then the covariance function $\Gamma(C,C') = E[X(C)X(C')]$ can be expressed in the form*

$$\sum_{i,j=1}^{N} f_i(C)f_j(C')h_{ij}(C,C'), \tag{5.5.15}$$

where the matrix $(h_{ij}(C,C'))$ is a Gramian, the entries of which are $(g_i,g_j), 1 \le i,j \le N$, and $h_{ij}(C,C')$ is a function of a pair $\{C,C'\}$.

Corollary 5.3 *If $N = 1$, then it is (simple) Markov.*

5.6 Euclidean free field

Euclidean free field has been discussed with the hope that it turns into a quantum field under the analytic continuation of the time variable from imaginary to real.

A Euclidean free field is a random field with parameter space $R \times R^d = R^{d+1}$, where a member of which is denoted by $(t,u), t \in R, u \in R^d, t$ being the time variable and u being the space variable.

Let $X(t,u)$ be a random field. If its characteristic functional is expressed in the form

$$C_\Gamma(\xi) = E[e^{i\langle x,\xi\rangle}] = \exp\left[-\frac{1}{2}\langle\xi,\Gamma\xi\rangle\right],$$

where the variable ξ is a member of some nuclear space E which is dense in $L^2(R^{d+1})$. The bilinear form $\langle\xi,\Gamma\xi\rangle$ is assumed to be a positive continuous bilinear form on $E \times E$.

By the Bochner Minlos theorem, there exists a unique probability measure μ_C on E^* such that

$$C_\Gamma(\xi) = \int_{E^*} e^{i\langle x,\xi\rangle} d\mu_C(x)$$

(see Appendix 1). The measure μ_C is Gaussian and has a covariance functional $\Gamma(\xi,\eta) = \langle \xi, \Gamma\eta\rangle$. We are thus given a random field $X((t,u),x), x \in E^*$, parameterized by a time-space variable $(t,u) \in R^{d+1}$.

Now it is an interesting question if the field $X(t,\omega)$ defines a quantum field by the so called Wick rotation $t \to -it$ or $t \to it$. To answer this question, it suffices to check the so called Osterwalder–Schrader axioms:

[OS-0.] Analyticity
[OS-1.] Euclidean invariance
[OS-3.] Reflection positivity (T-positivity)
[OS-4.] Ergodicity.

The details of the meaning of these axioms can be found in J. Glimm and A. Jaffer [13].

The only condition OS-3, Reflection positivity, is crucial. For this purpose, there is an important theorem.

Before stating the theorem, we recall the following definitions.

Definition 5.8 (OS-3) Let θ be the time reflection

$$\theta : \theta f(u,t) = f(u,-t).$$

If for any member Y of the algebra generated by the $e^{\langle X,\xi\rangle}, \xi \in E$,

$$\langle \theta Y, Y\rangle = \int \theta Y(x)\overline{Y(x)} d\mu \geq 0,$$

then we say that X satisfies reflection positivity.

Definition 5.9 A bilinear form Γ on $E \times E$ satisfies reflection positivity if for every $\xi \in E$ supported by positive time it holds that

$$\langle \theta\xi, P\xi\rangle \geq 0.$$

Theorem 5.10 *The Gaussian random field $X((t,u))$ satisfies reflection positivity if and only if Γ does, with the assumption that Γ commutes with θ.*

Proof. Assume that Γ is reflection positive, and set

$$M_{i,j} = C_\Gamma(\xi_i - \theta\xi_j) = C_\Gamma(\xi_i)C_\Gamma(\xi_j)\exp[\langle \theta\xi_j, \Gamma\xi_j\rangle]$$

is positive so is $N_{ij} = e^{R_{ij}}$.

Then M_{ij} has positive eigenvalues if and only if $N_{ij} = \exp[\langle \theta f_i, CF_j \rangle]$ is a positive matrix. By assumption that $R_{ij} = \langle \theta f_i, Cf_j \rangle$ is a positive matrix, follows the positivity of $N_{ij} = \exp[R_{ij}]$. This proves the theorem.

Now two examples that satisfy the reflection positivity are given in the following.

Example 5.4 The case R^1 (only time parameter).
Let $X(t)$ be a purely non-deterministic stationary Gaussian process. Then, $X(t)$ satisfies the reflection positivity if its covariance function $\gamma(h)$ is expressed in the form

$$\gamma(h) = \int_0^\infty \exp[-|h|\lambda] dm(\lambda),$$

where dm is a positive finite measure. See [83].

Example 5.5 Free Markov field
Consider the free Markov field in Nelson's sense of mass m on d-dimensional Euclidean space. There are good literature [60, 61], indeed they are heuristic, to show how to construct quantum fields from such Markov fields.

Remark 5.5 *Y. Okabe has extensively developed reflection positivity for Gaussian processes, see e.g. [64]. After this followed several papers to be noted.*

Chapter 6

Some Non-Gaussian Random Fields

We now come to the study of non-Gaussian random fields of two different kinds. One comes from the homogeneous chaos and the other is a functional of Poisson noise. In both cases the random fields are parameterized by a member C in a class $\mathbf{C}$; defined as in (5.2.1).

6.1 Fields of homogeneous chaos

First we take a random field expressed as a homogeneous chaos of white noise $x(u), u \in R^1$. To fix the idea let us consider the random field of homogeneous chaos of degree n. Namely, let $X(C), C \in \mathbf{C}$, be given by

$$X(C) = \int_{(C)} F(C; u_1, u_2, \ldots, u_n) : x(u_1)x(u_2) \cdots x(u_n) : du^n, \qquad (6.1.1)$$

where the kernel $F(C; u_1, u_2, \ldots, u_n)$ is a non-random function of C and $L^2(R^n)$-function of $(u_1 \cdot u_2, \ldots, u_n)$ vanishing outside of (C). Such a kernel is called a *Volterra kernel.* It is a causal representation in terms of homogeneous chaos.

For notational convenience the kernel is written simply as $F(C, \mathbf{u})$ where $\mathbf{u}$ denotes an n-dimensional vector. The factor $: x(u_1)x(u_2) \cdots x(u_n) :$ of the integrand is the Wick product of $x(u_j)$'s, which may be written, in response to the notation of the kernel, as $: x^{n\otimes} :$. Thus, $X(C)$ may be expressed in the form

$$X(C) = \int_{(C)} F(C; \mathbf{u}) : x^{\otimes n}(\mathbf{u}) : du^n. \qquad (6.1.2)$$

(For the Wick product : :, see Section 2.1.) The canonical property of the representation of the above representation can be defined in a similar manner to the case of a Gaussian process by replacing $\dot{B}(u)du$

with $: x^{n\otimes}(u) : du^n$. However, to make the matters simple, we give a definition of the canonical representation by using somewhat different terminology, although essentials are the same.

For a Borel set $U \subset R^n$, set

$$Z(U) = \int_U : x^{\otimes n}(\mathbf{u}) : du^n.$$

Then, $Z(U)$ defines an orthogonal random measure. For, the additivity in U is obvious. Also, finitely additive property is obvious. By using orthogonality of $Z(U_1)$ and $Z(U_2)$ for disjoint U_1 and U_2, we can prove countable additivity. This means $\{dZ(U)\}$ is an orthogonal random measure. With this understanding the integral (6.1.2) is considered as a stochastic integral with respect to $dZ(u)$.

Let $M(C)$ be the closed linear subspace of $L^2(\Omega, P)$ spanned by the $X(C')$'s with $(C') \subset (C)$. The projection operator down to $M(C)$ is denoted by $P_{M(C)}$.

Definition 6.1 A representation of a random field $X(C)$ given by the formula (6.1.2) is called *canonical* if the equality

$$P_{M(C')}X(C) = \int_{(C')} F(C, \mathbf{u}) : x^{n\otimes}(\mathbf{u}) : du^n$$

holds for any pair $\{C, C'\}$ with $C' < C$, and $F(C, \mathbf{u})$ is called a canonical kernel.

Proposition 6.1 *The canonical representation is unique, if it exists.*

Proof. The assertion is easily proved, so it is omitted.

Theorem 6.1 *A kernel function $F(C; \mathbf{u})$ is a canonical kernel if*

$$\int_{(C')} F(C', \mathbf{u})\varphi(\mathbf{u})du^n = 0 \quad for\ any\ C' < C$$

implies

$$\varphi(\mathbf{u}) = 0, \ \ a.e.\ in\ (C).$$

The proof is similar to the Gaussian case.

From the result in the paper [33] we can obtain the innovation $\{x(s), s \in C\}$ from $\delta X(C)$ and the $X(C')$'s with $C' < C$.

As for the general theory of the innovation, we shall discuss in Chapter 8 for general random fields.

Before closing this section a short note is given. Namely, we now briefly discuss the prediction theory for $X(C)$. Still it is assumed that $X(C)$ has the canonical representation with the canonical kernel $F(C;\mathbf{u})$.

First the best linear prediction will be given. For a fixed C let $X(C)$ be regarded as the present stage. Then, for C' with $C' < C$, the $X(C')$ is tacitly understood to be the past value and the $X(C'')$ with $C'' > C$ is the future value. Now assume the past values are known. The best linear predictor $X(C''|C)$ for the future value $X(C'')$ with $C'' > C$ is given by the projection of $X(C'')$ down to the linear space $M_C(X)$ spanned by the $X(C')$ with $C' \leq C$. Let P_M be the projection operator down to the subspace M as before. Then

$$X(C''|C) = P_{M_C(X)}X(C''). \tag{6.1.3}$$

Proposition 6.2 *Let $X(C)$ be given by* (6.1.2) *which is canonical. The best linear predictor is given by*

$$X(C''|C) = \int_{(C)} F(C'';\mathbf{u}) : x^{\otimes n}(\mathbf{u}) : du^n. \tag{6.1.4}$$

Proof. By using the canonical property of the kernel $F(C;\mathbf{u})$ it is proved that the linear space $M_C(X)$ is equal to the linear space spanned by the $x^{n\otimes}(\mathbf{u}), u \in (C)$. Hence, holds the equality shown in the theorem.

We now obtain the nonlinear predictor, which is usually called simply the predictor. In the present case, namely in the case where the random field is given by the homogeneous chaos that is monomial, the following theorem can be proved.

Theorem 6.2 *The best nonlinear prediction for the field $X(C)$ given by* (6.1.1) *is the same as the best linear predictor.*

Proof. Under the same situation of the present, past and future, the (best) nonlinear prediction is the conditional expectation $E[X(C'')|\mathbf{B}(C)]$, where $\mathbf{B}(C)$ is the sigma-field of events determined by $X(C'), C' < C$. There is an important fact; namely, the conditional expectation is a nonlinear function of the $X(C')$ with $C' < C''$ (note that no more Gaussian case). It is therefore a nonlinear predictor in the weak sense.

6.2 Multiple Markov properties of homogeneous chaos $X(C)$

The idea of introducing the multiple Markov properties is the same as in the Gaussian process and Gaussian random field.

Let $X(C)$ be as in the last section. Since it is a linear functional of the homogeneous chaos, the multiple Markov property can be defined in a similar manner to the case of Gaussian process by using the conditional expectations. To make sure, we give

Definition 6.2 Let $X(C)$ be given by

$$X(C) = \int_{(C)^n} F(C, \mathbf{u}) : x^{\otimes n}(\mathbf{u}) : du^n \tag{6.2.1}$$

with a canonical kernel F. For any choice of C_i's such that $C_0 \leq C_1 < \cdots < C_N < C_{N+1}$,

(1) $E[X(C_i)|\mathbf{B}_{C_0}(X)], i = 1, 2, \ldots, N$, are linearly independent and

(2) $E[X(C_i)|\mathbf{B}_{C_0}(X)], i = 1, 2, \ldots, N+1$ are linearly dependent

then, $X(C)$ is said to be *N-ple Markov* homogeneous chaos random field in the weak sense.

One can compare this definition with that for $X(t)$ (see Section 5.5) to recognize a plausibility of the present definition,

Theorem 6.3 *If a random field $X(C)$ of homogeneous chaos is N-ple Markov, then its canonical kernel is a Goursat kernel of order N.*

Proof. For proof we only note that the conditional expectation is a non-linear function of the known values, unlike Gaussian case. For the rest of the proof we can follow the method given in Section 5.5B.

Corollary 6.1 *The predictor of an N-ple Markov random field of homogeneous chaos is computable. More precisely, the best predictor is a linear combination of the random variables obtained from the values of the past.*

Here we note that the multiple Markov properties indicate not only the way of dependence, but also suggest computability of the best predictor by using finite number of basic vectors.

Remark 6.1 *The notion of the canonical kernel can be used for non-Gaussian case, if a random field or a stochastic process is expressed as an integral of such a kernel with respect to some random measure (see e.g. Accardi–Hida–Si Si* [2]*).*

6.3 The Poisson case

Before we come to a field formed by a Poisson noise, we have to remind the general innovation $Y(t)$ of a process $X(t)$. One can see that, in favourable cases, there is an additive process $Z(t)$ such that its derivative $\dot{Z}(t)$ is equal to the $Y(t)$, since the collection $\{Y(t)\}$ is an independent system. There is tacitly assumed that, in the system, there is no random function which is singular in t.

There is the Lévy decomposition of an additive process. If $Z(t)$ has stationary independent increments, then except trivial component the $Z(t)$ involves a compound Poisson process and a Brownian motion up to constant. With this remark in mind we proceed to the Poisson case.

After Brownian motion comes another kind of elemental additive process which is to be the Poisson process denoted by $P(t), t \geq 0$. Taking its time derivative $\dot{P}(t)$ we have a *Poisson noise* as is mentioned in Section 3.1. More generally, we may consider a compound Poisson noise, however to make the matters simple, we take just a Poisson noise, in fact the result in this case can be generalized to the case of compound Poisson case in a routine manner.

Also, for simplicity, we start with a case of processes, later a generalization to a field will be discussed.

A Poisson noise is a generalized stationary stochastic process with independent value at every point. For convenience we may assume that t runs through the whole real line. In fact, it is easy to define such a noise. The characteristic functional of the centered Poisson noise is of the form (3.1.2).

There is the associated measure space (E^*, μ_P), and the Hilbert space $L^2(E^*, \mu_P) = (L^2)_P$ is defined.

Many results of the analysis on $(L^2)_P$ have been obtained, however most of them have been studied by analogy with the Gaussian case or its modifications, as far as the construction of the space of generalized functionals. Here we only note that the $(L^2)_P$ admits the direct sum decomposition of the form as in Section 3.1:

$$(L^2)_P = \bigoplus_n H_{P,n}.$$

Let $P(t)$ denote a Poisson process as before. The subspace $H_{P,n}$ is spanned by Poisson noise functionals defined by the product of the Poisson–Charlier polynomials of degree n as follows. Let $p(x; \lambda)$ be the

Poisson distribution with intensity $\lambda > 0$:

$$p(x;\lambda) = \frac{\lambda^x}{x!}e^{-\lambda}, \quad x = 0, 1, 2, \ldots.$$

Then

$$p_n(x;\lambda) = \frac{\lambda^{n/2}}{n}(-1)^n \frac{\Delta^n p(x;\lambda)}{p(x;\lambda)},$$

where Δ is the difference operator: $\Delta f(x) = f(x) - f(x-1)$.

The polynomials $p_n(x;\lambda), n = 0, 1, 2, \ldots,$ form an orthogonal system in the following sense:

$$\sum_{x=0}^{\infty} p_n(x;\lambda)p_m(x;\lambda)p(x;\lambda) = \delta_{n,m}.$$

We take various time interval I with $|I| = \lambda$ and form a subspace spanned by the Poisson–Charlier polynomials $\prod_{n_k}(P(I_k))$ of degree $n = \sum n_k$ in the variables $P(I_k)$'s, where $P(I) = P(b) - P(a)$ for $I = [a, b]$. Then we have the subspace $H_{p,n}$.

By the way of construction and the orthogonality of the Poisson–Charlier polynomials it is easy to see the orthogonality of the subspaces $H_{P,n}$ for different n, where the addition formula of the Poisson–Charlier polynomials is used.

$$\begin{aligned}\left(\frac{(a+b)^n}{n!}\right)^{1/2} & p_n(x+y+1, a+b) \\ &= \sum_{m=0}^{n}\left(\frac{a^m b^{n-m}}{m!(n-m)!}\right)^{1/2} p_m(x,a)p_{n-m}(y,b). \qquad (6.3.1)\end{aligned}$$

Concerning a stochastic process there might occur a misunderstanding if it is expressed as a functionals of Poisson noise, even in the case of linear functional. The following example would illustrate this fact.

Let a stochastic process $X(t)$ be given by an integral

$$X(t) = \int_a^t F(t,u)\dot{P}(u)du, \quad t \geq a,$$

where $F(t,u)$ is continuous in (t,u), with $u \leq t$.

It seems to be simply a linear functional of $P(t)$, however there are two ways of understanding the meaning of the integral in such a way that

(i) the integral is defined in the Hilbert space by taking $\dot{P}(t)dt$ to be a random measure, and the stochastic integral is defined. Like the homogeneous chaos, multiple integral can also be defined,
(ii) another integral is understood as a continuous bilinear form of a test function $F(t,\cdot)$ and a sample function of $\dot{P}(t)$ (the path-wise integral (see Gel'fand [10], 1955)). This can be done if the kernel is a smooth function of u over the interval $[a,t]$ since a sample function of $\dot{P}(t)$ is a generalized function of t.

Assume that $F(t,t)$ never vanishes and that it is not a canonical kernel, that is, it is not a kernel function of an invertible integral operator. Then, we can claim that for the integral in the first sense $X(t)$ has less information compared to $P(t)$. Because there is a linear function of $P(s), s \le t$ which is orthogonal to $X(s), s \le t$. On the other hand, if $X(t)$ is defined in the second sense, and if it is modified to be the right continuous in t, then we can prove the following proposition. Before a rigorous assertion is stated, we provide a notation:

$$\mathbf{B}_t(\dot{P}) = \bigwedge_{\varepsilon>0} \bigvee_{\text{supp}(\xi)\supset[a,t+\varepsilon)} \mathbf{B}(\dot{P}(\xi)), \tag{6.3.2}$$

where $\dot{P}(\xi) = \langle \xi, \dot{P} \rangle$.

Proposition 6.3 *Under the assumptions stated above, if the $X(t)$ above is defined sample function-wise, we have the following equality for sigma-fields:*

$$\mathbf{B}_t(X) = \mathbf{B}_t(\dot{P}), \quad t \ge 0.$$

Proof. By assumption it is easy to see that $X(t)$ and $P(t)$ share the jump points, which means that the information is fully transferred from $P(t)$ to $X(t)$. This proves the equality.

The above argument tells us that we are led to introduce a space $(\mathbf{P})$ of random variables that comes from separable stochastic processes for which existence of variance can not be expected. This sounds to be a vague statement, however we can rigorously define it by using a Lebesgue space without atoms, and others. There the topology is defined by either the almost sure convergent or the convergence in probability, and there is no need to think of mean square topology. On the space $(\mathbf{P})$ filtering and prediction for strictly stationary process can naturally be discussed. For some related idea we may refer to the literatures [17] and [18], where one can see further profound idea of N. Wiener [93].

It is almost straightforward to come to an introduction to a multi-dimensional parameter Poisson noise, denoted by $\{V(u)\}$, which is a multi-dimensional parameter generalization of $\{\dot{P}(t)\}$.

A multi-dimensional, say d-dimensional, parameter Poisson noise $\{V(u), u \in R^d\}$ is a generalized stochastic process parameterized by R^d such that its characteristic functional $C_P(\xi) = E[\exp[i\langle V, \xi\rangle]$, $\xi \in E$ is given by

$$C_P(\xi) = \exp\left[\lambda \int_{R^d} (e^{i\xi(u)} - 1) du^d\right], \quad \xi \in E. \tag{6.3.3}$$

The existence is guaranteed by the Bochner–Minlos theorem (see [18]) where E is taken to be a nuclear subspace of $L^2(R^d)$.

The probability distribution of V is denoted by μ_P over $(E^*, \mathcal{B})$, $\mathcal{B}$ being a *sigma*-field generated by cylinder subsets of E^*. The space $(E^*, \mathcal{B}, \mu_P)$ is viewed as a realization of a d-dimensional Poisson noise, and hence μ_P almost all $x \in E^*$ are considered as sample functions of V.

Remind the discussion in Section 3.5 for the idea behind the construction.

Theorem 6.4 *Let a random field $X(C)$ parameterized by a contour C be given by a stochastic integral*

$$X(C) = \int_{(C)} G(C, u) V(u) du,$$

where the kernel $G(C, u)$ is continuous in (C, u). Assume that $G(C, s)$ never vanishes on C for every C. Then, the $V(u)$ is the innovation.

Proof. The variation $\delta X(C)$ exists and it involves the term

$$\int_C G(C, s) \delta n(s) V(s) ds,$$

where $\{\delta n(s)\}$ determines the variation δC of C. Here the same technique is used as in the case of [57], so that the values $V(s), s \in C$, are determined by taking various δC's. This shows that the $V(s)$ is obtained by the $X(C)$ according to the infinitesimal change of C. Hence $V(s)$ is the innovation.

Here is an important remark. In the Poisson case one can see much similarity on getting the innovation to the case of a representation of a

Gaussian process provided the integral discussed just above is defined as a stochastic integral with respect to a random measure $V(u)du$ or $V(s)ds$. However, if one is permitted to use some nonlinear operations acting on sample functions, it is possible even to form the innovation from a non-canonical representation.

The situation is the same as in the one-dimensional parameter case.

6.4 Poisson noise functionals

Orthogonal polynomials (in fact, products of them) in Poisson noise $V(u)$ of degree n span a space of the *discrete chaos* of degree n, and it is denoted by $H_{P,n}$. The space (L^2_P) admits the direct sum decomposition (see Section 3.1):

$$(L^2)_P = \bigoplus_n H_{P,n}.$$

We are now ready to come to a compound Poisson process, which is a more general innovation, the second order moment may not exist, so that we have to come to the space $(\mathbf{P})$, defined as a linear space of random variables measurable with respect to some separable sigma-field. The topology is defined by the almost sure convergence or by convergence in probability.

Some basic properties of compound Poisson noise have been discussed in Section 3.8, so that just brief notes will be added.

The Lévy decomposition of an additive process $Z(t), t \geq 0$, with which we are now concerned, is expressed in the form

$$Z(t) = \int \left(uP_{du}(t) - \frac{tu}{1+u^2} dn(u) \right) + \sigma B(t),$$

where $Z(t)$ is assumed to have stationary increments and is continuous in probability. In the above formula, the variable u stands for the amount of jump of the component Poisson process, and $P_{du}(t)$ is a random measure of the set of Poisson processes, and where $dn(u)$ is the Lévy measure supported by $(-\infty, 0) \cup (0, \infty)$ such that

$$\int \frac{u^2}{1+u^2} dn(u) < \infty.$$

The decomposition of a compound Poisson process into the elemental Poisson processes, each of which has different jump, can be carried out in the space $(\mathbf{P})$ with the use of the quasi-convergence (see [46, Chapter V]). The tensor product of L^2-spaces associated with those elemental Poisson processes can be formulated by the well-known theory of Hilbert space. While, in the space (**P**), functionals of sample function can be treated. With this fact in mind we are ready to discuss the analysis acting on sample functions of a compound Poisson process.

To have a generalization of Theorem 6.4 in the previous section to the case of compound Poisson noise is not difficult in a formal way without paying much extra attention. However, we wish to pause at this moment to consider carefully about how to find a jump point of $Z(t)$ with the height u chosen in advance. This question, so-to-speak the jump finding problem, is heavily depending on the computability or measurement problem. Questions related to this problem shall be discussed in our forth coming paper.

6.5 Random fields as Poisson noise functionals

(1) A Brownian motion and each Poisson process of the component of a compound Poisson process seem to be elemental and the two are entirely different from each other. Indeed, this is true in a sense. On the other hand, there is another aspect. We know that the inverse function of the Maximum of a Brownian motion is a stable process, which is a compound Poisson process (see [46, Chapter VI]). A Poisson process may therefore come from a Brownian motion in (**P**). It should be noted that it does certainly not by the L^2 method in the ordinary sense. The same situation can be seen in the analysis of random fields.

Also, in terms of the probability distribution, it is quite interesting to note that some generalized (Gaussian) white noise functional has the same distribution as that of a Poisson white noise (see in Cochran–Kuo–Sengupta, IDAQP Vol. 1, 1998). There arises a question on how to find concrete operations (variational calculus may be involved there) acting on the sample functions of $\dot{B}(t)$'s to have a Poisson noise. We need some more examples to propose a problem to give a good interpretation to such phenomena.

(2) In the space (P) we can emphasize upon the importance of "quasi-convergence". This plays similar role to the renormalization in Gaussian case to observe a generalized functionals in $(S)^*$. Note that the existence of variance or any order of moments is not assumed. Such a freedom is more convenient in the study of random field.

(3) Unlike the Gaussian case, there is only a very poor group of invariance, the action of which keeps the Poisson noise measure μ_P invariant. This

is true for the case a general compound Poisson noise. In fact, the shift and the reflection with respect to the origin form a group which keeps the measure in question invariant. So, harmonic analysis arising from this small group will do not work so effectively, although we can say something because the shift is very important in many ways. This causes some disadvantages when we make actual computation of the variation of random fields $X(C)$ by taking deformations of C using the group acting on the parameter set.

However, if we come to a stable process, then we can see some more invariance, mostly based on the dilation of the parameter. Such an observation gives us some naive interpretation on the structure of the associated measure introduced on the space of generalized functions.

Chapter 7

Variational Calculus For Random Fields

A most successful method to analyze random fields indexed by a manifold C is the variational calculus. We propose an analytic method of the approach to this topic.

7.1 Generalized white noise functionals and random fields

Random fields which are going to be discussed in this chapter are functionals of white noise, either Gaussian or of Poisson type.

There are two topics, the significance of which are emphasized; namely one is the efficient use of the classical theory of variational calculus and the other is the restriction of the parameter of the white noise. As for the latter, given a field formed by a functional of multi-parameter noise, its variation usually involves a functional of a lower dimensional parameter white noise. Unlike the case for non-random fields, our calculus needs to consider a restriction of parameter of white noise. How to understand the restriction of parameter of white noise has been explained in Sections 2.2 and 3.5, however another method that directly connected to the variation of a random field is illustrated below.

To make the matters simple, we first review the Gaussian case. Let (E^*, μ) be a white noise with R^d-parameter; $E \subset L^2(R^d)$ as a background. The complex Hilbert space $(L^2) = L^2(E^*, \mu)$ is introduced in the usual manner, and we have the Fock space:

$$(L^2) = \bigoplus_{0}^{\infty} H_n.$$

According to the integral representation or by the S-transform we are given an isomorphism

$$H_n \cong \widehat{L^2(R^{nd})},$$

where $\widehat{L^2(R^{nd})}$ is the subspace of $L^2(R^{nd})$ involving functions which are symmetric in the d-dimensional vectors.

Also, in the same manner as in the one-dimensional parameter case, the space $(S)^*$ of generalized white noise functionals will be introduced, by using the second quantization method for the operator

$$A = -\Delta_d + \sum u_j^2 + 1,$$

where Δ_d is the d-dimensional Laplacian operator.

Thus we have a so-called Gel'fand triple:

$$(S) \subset (L^2) \subset (S)^*.$$

In the space $(S)^*$ there is a subspace H_n^- which is a natural extension of the homogeneous chaos H_n subordinated to the extension of (L^2) to $(S)^*$. Take, in particular, the subspace H_1^-, any member of which is expressed as a bilinear form $\langle x, f \rangle$, where f is in the Sobolev space over R^d of degree $-(d+1)/2$. The same for H_n^-.

7.2 Restriction of parameter (continued)

We are now ready to give an interpretation of restriction of the parameter to a suitable $(d-1)$-dimensional manifold C. As C has been specified to be convex C^∞-manifold, so that $\langle x, \xi|_C \rangle$ makes sense, where $\xi|_C$ is such a generalized function that

$$\langle \xi|_C, \eta \rangle = \int_C \xi(s)\eta(s)ds,$$

where ds is the line element of C. Thus,

$$C_C(\xi) = \int_{E^*} \exp[i\langle x, \xi|_C \rangle] d\mu(x),$$

which is expressed in the form

$$C_C(\xi|_C) = \exp\left[-\frac{1}{2}\int_C |\xi(s)|^2 ds\right].$$

Noting that $\xi|_C$ is a test function defined on C. This fact follows the choice of the manifold C. Hence we are given a white noise parameterized by a point s in C.

Our next question is how to apply the classical theory of functional analysis to the expression of the variational formula for a random field $X(C)$ which is assumed to be a linear functional of Gaussian white noise or of homogeneous chaos, or even of Poisson noise.

So we have to consider all these cases:

(i) Gaussian case
(ii) Homogeneous chaos
(iii) Poisson noise.

As for (i) we have discussed in Section 2.2 and for (iii) detailed consideration has been given in Section 3.5. Only for the case (ii) we need interpretation and it is as follows.

The random measure defined by the Wick product $: x(u_1) \cdots x(u_n) :$ can be restricted to a random measure on a surface by taking integrand to be generalized function which is supported by the surface.

7.3 Variational formula for $X(C)$

We observe, in this section, the variation of $X(C)$ when C bears an infinitesimal deformation.

The parameter C is always assumed to be a member of the class $\mathbf{C}$ as before.

Let $X(C)$ be an (L^2)-functional of white noise $x(u)$. Then we can apply the S-transform to have a functional $U(C, \xi)$. Let ξ be fixed for a moment and use the simple notation $U(C)$ or $S(X(C))$.

The variation of $X(C)$ is therefore defined by

$$\delta X(C) = S^{-1}\delta(S(X(C))(\xi)) = S^{-1}(\delta U(C)). \tag{7.3.1}$$

Variational calculus for a random field $X(C)$ can be reduced to that for the associated U-functionals. The details of the variational calculus for non-random functionals including functional $U(C)$ are given in the Appendix 2. Thus, so far as the calculus on (L^2) is concerned, we can see that everything related to the calculus is done by the operations acting on U-functionals.

Now one may ask the main purpose of the variational calculus for $X(C)$. The answer is two-fold.

(1) The stochastic variational equation involving the innovation (see the equation in Section 1.3. for the idea) can characterize the probabilistic structure of $X(C)$. To solve the equation is an attributive work.
(2) The second one is to obtain the innovation. The relationship with other notions such as the past values, C etc., is given by the variational equation.

More concrete description will be seen in the examples which enjoy their own characteristic meaning.

Particular interest may be seen in the examples where generalized white noise functionals appear. Singularity is involved in the integral representation of those functionals. In the similar meaning to that for the non-random case discussed in Appendix 2, the singularity on the diagonal has been well justified. Whilst, it has discovered recently that singularity appearing off-diagonal should have another important meaning corresponding to the quantum mechanical phenomena.

We start with a simple example to illustrate the idea. Let $\mathbf{C}$ be the system given in Section 5.2 and let $Y(C), C \in \mathbf{C}$ be defined by

$$Y(C) = \int_{(C)} g(u)x(u)du^d, \tag{7.3.2}$$

where (C) is the domain enclosed by C, and where g is a continuous function. Then, it is a Gaussian random field, and we have a causal representation of a Gaussian random field $Y(C)$ in terms of white noise $x(u), u \in R^d$.

Before we come to the variational equation, we have to prepare some particular properties which never appear in non-random case. To make the matters simple, the parameter space is taken to be R^2.

An ordinary linear functional X of two-dimensional parameter white noise is expressed in the form

$$X = \int f(u)x(u)du^2,$$

where f is a square integrable function. The S-transform $U(\xi)$ of X gives us

$$U(\xi) = \int f(u)\xi(u)du^2.$$

Now we may consider the case where f is a generalized function, say a member of the Sobolev space of order $-3/2$. Then, X is a generalized white noise functional in H_1^{-1}. In particular, if f represents a curve C, then we may have $f(u,v)\delta(v-\psi(u))$ locally, as the kernel function of X, where ψ is a function that determines the curve C. For notation, let it be denoted by X_C. If we are allowed to introduce a term, analogous to the linear parameter case, a generalized field with independent values at every C, the above X_C, is just fitting for such a field. This can be justified in the following manner. Take a continuously increasing family $\mathbf{C}$ of contours C and assume that $\mathbf{C}$ is homeomorphic to the family of straight lines in the plane. Then, the assertion in Section 4.2 can be applied to the case of the consistent family of random fields with independent values at every straight line. Hence, we can form an innovation for the field for X. Summing up we can claim the following proposition.

Proposition 7.1 *The variation of $Y(C)$ defined by (7.3.2) is expressed in the form*

$$\delta Y(C) = \int_C g(s)x(s)\delta n(s)ds, \tag{7.3.3}$$

where ds is the line element over C. The functional derivative is, therefore, given by

$$\frac{\delta Y(C)}{\delta n}(s) = g(s)x(s), \quad s \in C, \tag{7.3.4}$$

from which the innovation $x(s)$ is obtained.

Proof. Take the S-transform of $Y(C)$ to have $U(\xi) = \int_{(C)} g(u)\xi(u)du^d$. Take the variation and send it back to the functional of $x(s)$ to have $\delta Y(C)$. The functional derivative of $U(\xi)$ is $g(s)\xi(s)$ which corresponds to $g(s)x(s)$, an additive Gaussian process $X(t) = \int_0^t g(s)x(s)ds$ is defined, where t is taken to be the arc length of C. Hence, we can form the innovation $x(s)$ of $X(t)$, which is eventually the innovation of $Y(C)$. Then, it is proved, as is noted above. Thus the consistent family of innovations $\{x(s), s \in C\}$ defines the innovation of the whole parameter set. Thus, the assertion is proved.

Proposition 7.2 *Let a functional derivative of $Y(C)$ is given by the formula* (7.3.4), *where g is the restriction to C of some continuous function defined on R^2. Then, there exists a solution of* (7.3.4) *given by* (7.3.2). *The solution is uniquely determined with the initial data.*

Proof. Let g_C be the restriction of g to the domain (C). Then, a stochastic bilinear form $\langle x, g_C\rangle$ can be given. It is nothing but the Gaussian random field $Y(C)$ expressed in the form (7.3.2). Uniqueness comes from the fact that if the functional derivative is identically 0, then only a constant field is acceptable.

The $Y(C)$ discussed above is a random field analogue of an additive Gaussian process with parameter t or equivalently an analogue of Gaussian martingale. It is proved that a Gaussian martingale parameterized by C can be expressed like $Y(C)$ in (7.3.2).

The following example is concerned with a Markov field.

Let a Gaussian random field $X(C)$ be given. Assumed that $X(C)$ has a causal representation in terms of $x(u)$ and that $E(X(C)) = 0$. Further assume that it is a Markov field. Then by Theorem 5.8, there is a function $f(C)$ such that

$$X(C) = f(C)Y(C),$$

where $Y(C)$ is a martingale. Note that our definition of Markov field excludes the case of 0-ple Markov field.

Theorem 7.1 *Let $X(C)$ be a Gaussian Markov field with the expression $X(C) = f(C)Y(C)$, where $Y(C)$ is a martingale. A Borel measure σ is uniquely defined by the variance $\sigma((C)) = E(Y(C)^2)$. If the measure σ is absolutely continuous with respect to the Lebesgue measure, then*

(1) *the innovation $x(u)$ of $X(C)$ can be formed, and*
(2) *the canonical representation of $X(C)$ is expressed in terms of the innovation $x(u)$ in the form*

$$X(C) = f(C) \int_{(C)} g(u)x(u)du^d.$$

Proof. Note that the Markov property implies that $f(C) \neq 0$. It is easy to see that the measure σ is uniquely defined. Since the measure σ is assumed to be absolutely continuous with respect to the Lebesgue measure, one can find a function h such that

$$\sigma(B) = \int_B h(u)du^d,$$

and h never vanishes on any Borel set B of positive Lebesgue measure, in particular for every (C).

On the other hand, the variation $\delta Y(C) = \int_C g(s)\delta n(s)x(s)ds$ gives $g(s)x(s)$, sine δn can be chosen arbitrary. But we know $g(s)$ never vanishes since $g(s)^2 = h(s)$ which never vanishes almost everywhere. The rest of the proof is now obvious.

7.4 Variational equation

Let C be represented by a C^∞-function η. Then variation of η defines that of C. In what follows ξ is omitted, since it is arbitrary and it is fixed. So we consider the variational equation in the case where δU can be written as

$$\delta U(\eta) = \int_C f(\eta, U, s)\delta\eta(s)ds, \tag{7.4.1}$$

where $f(\eta, U, s)$ is continuous in three variables, satisfying the Lipshitz condition in U and has Fréchet derivative in η in the $C^\infty(S^{n-1})$-topology and has partial derivative in U. The equation of the form (7.4.1) is called a Volterra form. Assume that the Fréchet derivative and the partial derivative are continuous in all variables.

Thus the Fréchet derivative is given by

$$\frac{\delta U(\eta)}{\delta\eta(s)} = f(\eta, U, s). \tag{7.4.2}$$

P. Lévy has given the integrability condition formally as

$$\delta\delta_1 U = \delta_1\delta U. \tag{7.4.3}$$

We now give the rephrasement of the above integrability condition in an explicit form as is expressed in the following theorem, assuming analytic conditions on f as above. Existence of the solution to (7.4.1) will be discussed in the next section.

Theorem 7.2 *The integrability condition for the variational equation is given by*

$$\frac{\delta f(\eta, U, s)}{\delta\eta(t)} + \frac{\partial f(\eta, U, s)}{\partial U} f(\eta, U, t) = \frac{\delta f(\eta, U, t)}{\delta\eta(s)} + \frac{\partial f(\eta, U, t)}{\partial U} f(\eta, U, s). \tag{7.4.4}$$

Proof. According to the general integrability condition for the equation (7.4.1), we must have

$$\frac{\delta}{\delta\eta(t)}\frac{\delta U(\eta)}{\delta\eta(s)} = \frac{\delta}{\delta\eta(s)}\frac{\delta U(\eta)}{\delta\eta(t)} \tag{7.4.5}$$

in terms of Fréchet derivative (see [47] 2éme Part.). By noting that $\frac{\delta U}{\delta\eta(s)}$ is a function of the variable $\eta(t)$ and U (which is a function of $\eta(t)$'s), the left hand side of the above equation gives

$$\frac{\delta f(\eta, U, s)}{\delta\eta(t)} + \frac{\partial f(\eta, U, s)}{\partial U} f(\eta, U, t)$$

and the right hand side of the above equation gives

$$\frac{\delta f(\eta, U, t)}{\delta\eta(s)} + \frac{\partial f(\eta, U, t)}{\partial U} f(\eta, U, s).$$

Thus we have proved the assertion.

Remark 7.1 *The formula* (7.4.4) *does not depend on the choice of analytic expression of C in terms of η.*

Corollary 7.1 *If f is independent of η, then the equation* (7.4.2) *can be rephrased by*

$$f(U, s) = g(U)h(s). \tag{7.4.6}$$

Note. The f, in Corollary 7.1, satisfies (7.4.4) and the integrability condition holds.

Example 7.1 Consider a variational equation

$$\delta U(\eta) = U \int_0^{2\pi}\int_0^{2\pi} 2F(u, s)\eta(u)\delta\eta(s)du\, ds.$$

Then

$$\frac{\delta U(\eta)}{\delta\eta(s)} = f(\eta, U, s) = U \int_0^{2\pi} 2F(u, s)\eta(u)ds.$$

It can be easily verified that f satisfies the integrability condition, provided F is symmetric. Actually the solution is

$$U(\eta) = \exp\left[\int_0^{2\pi}\int_0^{2\pi} F(u, v)\eta(u)\eta(v)du\, dv\right].$$

Example 7.2 Consider a variational equation

$$\delta U(\eta) = U \int_0^{\infty} 2g(t)\eta(t)\delta\eta(t)dt.$$

Thus

$$f(\eta, U, s) = 2g(s)\eta(s)U.$$

It is easy to see that this f satisfies the integrability condition. The solution is the Gauss kernel, namely

$$U(\eta) = \exp\left[\int g(t)\eta(t)^2 dt\right].$$

Example 7.3 Set

$$\frac{\delta U(\eta)}{\delta \eta(s)} = f(\eta, U, s) = \int_0^s \eta(u)du; \quad 0 < s < 2\pi,$$

then the equation fails to satisfy the integrability condition. Thus there is no solution to the variational equation (7.4.1) with the f given in this example.

Example 7.4 Consider the Tomonaga–Schwinger equation

$$i\frac{\delta \Psi(C)}{\delta C_s} = H_s(C)\Psi(C),$$

where $H_s(C)$ is an operator and Ψ stands for the wave field depending on the surface C. We may formulate the equation in the white noise frame work by taking $\Psi(C) = X(C)$. Having applied S-transform technique and others, we are finally led to see the integrability conditions yields the Tomonaga's integrability condition

$$\frac{\delta H_s(C)}{\delta C_t} - \frac{\delta H_t(C)}{\delta C_s} = i[H_s(C)H_t(C)].$$

Indeed, this is a guiding equation, although we should study much concerning mathematical questions. It is noted that this equation was one of the motivations of our study on random field. For details, see Section 10.7.

7.5 Existence theorem for a variational equation

Our aim, remind again, is to establish the variational equation for a random field $X(C)$, and to discuss the solution to the variational equation. We are mainly interested in the random field which are functional of white noise or Poisson noise. There the S-transform or U-transform, respectively, carries $X(C)$ to a functional $U(C, \xi)$ of $\xi \in E$, so that our problem turns into the variational calculus for non-random functionals with variable C, where C runs through the class $\mathbf{C}$.

The variation of $U(C)$ is defined in the following way. Let C be the closed domain enclosed by the ovaloid C, and let $C < C'$ mean $(C) \subset (C')$.

An infinitesimal deformation of C now means that C expands a little bit outward to have $C + \delta C$. The distance $\rho(C, C + \delta C)$ from C to $C + \delta C$ is measured by $\sup_s \delta n(s)$, where $\delta n(s)$ is the distance along the normal from a point $a(s) \in C$ to $C+\delta C$, s being the parameter of C. This is in agreement with the topology introduced to C by using the Euclidean metric.

Now assume that U satisfies

$$U(C + \delta C) - U(C) = \delta U(C) + o(\rho),$$

where $\rho = \rho(C, C+\delta C)$, and where $\delta U(C)$ is a continuous linear functional in some neighbourhood (in the ρ-topology) of C.

By assumption we are given a Volterra form of $\delta U(C)$. Namely, it is expressed in the form

$$\delta U(C) = \int_C U'(C, s)\delta n(s) ds. \tag{7.5.1}$$

The $(d-1)$-dimensional parameter s for a representation of C is taken to be that is determined by the Euclidean metric in R^d, and ds is the surface element over C.

The kernel function $U'(C, s)$ is the derivative of $U(C)$ in the variable C. If $U'(C, s)$ is an image of some generalized white noise functional under S, then $(S^{-1}U'(C, s))(x)$ is the functional derivative of $X(C)$, and is denoted by $\frac{\delta X(C)}{\delta C}(s)$.

We now come to the solution of a stochastic variational equation of the form (7.5.1) when $U'(C, s)$ is given with initial conditions.

By our usual method, we rewrite it having applied the S-transform. Hence we have a variational equation of the form

$$\delta U(C) = \int_C f(C, U, s)\delta n(s) ds, \tag{7.5.2}$$

where C may be replaced by a function that represents the ovaloid C.

Consider the particular case where C is diffeomorphic to a unit circle S^1, so that $U(C)$ can be expressed as a function of ξ, U and t:

$$U(C) = U(\xi, U, t), \quad \xi \in E, \;\; t \in I = [0, 1],$$

where E being a nuclear subspace of $L^2(I)$.

Hence the variational equation (7.5.2) may be expressed in the form

$$\delta U = \int_I f(\xi, U, \tau)\delta\xi(\tau) d\tau. \tag{7.5.3}$$

Having been suggested by P. Lévy [47], we now discuss how to solve this equation with the initial condition $U_0 = U(\xi_0)$. For a rigourous argument, several assumptions are recommended such as (A.1), (A.2) and (A.3):

(A.1) $f(\xi, U, s)$ is continuous in the three variables $(\xi, U, s) \in E \times R \times I$ and satisfies the integrability condition,

(A.2) In some neighbourhood of U_0, e.g. $|U - U_0| < C$, $|V - U_0| < C$, there exist constants M and K such that

(i) $\int_I f(\xi, U, s)^2 ds < M^2$,
(ii) $\int_I [f(\xi, U, s) - f(\xi, V, s)]^2 ds < K^2(U - V)^2$.

First, we discuss the existence of the solution with variable ξ running through an arrow, so that we can write

(A.3) $\xi = \xi_0 + \lambda\eta, \quad \|\eta\| = 1, \ \lambda \geq 0$, where $\| \ \|$ is the $L^2(I)$-norm.

Theorem 7.3 *Under the assumptions* (A.1) $\sim$ (A.3), *the solution of the variational equation* (7.5.3) *exists and unique.*

Proof. Set $\varphi_0 = U_0$. Define $\varphi_p(\lambda)$ by induction

$$\varphi_p(\lambda) - \varphi_0(\lambda) = \int_0^\lambda d\nu \int_I f(\xi_0 + \nu\eta, \varphi_{p-1}(\nu), s)\eta(s)ds. \tag{7.5.4}$$

Then, we have

$$\begin{aligned} &\varphi_p(\lambda) - \varphi_{p-1}(\lambda) \\ &\quad = \int_0^\lambda d\nu \int_I (f(\xi_0 + \nu\eta, \varphi_{p-1}(\nu), s) - f(\xi_0 + \nu\eta, \varphi_{p-2}(\nu), s))\eta(s)ds. \end{aligned} \tag{7.5.5}$$

Assume that

$$|\varphi_i(\nu) - \varphi_0(\nu)| < C, \quad i = 1, 2, \ldots, p-1; \ \nu < \lambda. \tag{7.5.6}$$

Then, by (A.2)(ii),

$$\varphi_p(\lambda) - \varphi_{p-1}(\lambda) < \int_0^\lambda K|\varphi_{p-1}(\nu) - \varphi_{p-2}(\nu)|d\nu. \tag{7.5.7}$$

This proves successively

$$\begin{aligned} &|\varphi_1(\lambda) - \varphi_0(\lambda)| < M\lambda \\ &|\varphi_i(\lambda) - \varphi_{i-1}(\lambda)| < K^{i-1}M\frac{\lambda^i}{i!}, \quad i = 2, \ldots, p, \end{aligned}$$

so that

$$|\varphi_p(\lambda) - \varphi_0(\lambda)| < \frac{M}{K} \sum_{i=1}^{p} \frac{K^i \lambda^i}{i!} \leq \frac{M}{K}(e^{k\lambda} - 1).$$

The λ can be chosen small enough to hold $\frac{M}{K}(e^{k\lambda} - 1) < C$, and hence (7.5.6) holds for every p. Applying (7.5.7) successively, we have

$$|\varphi_p(\lambda) - \varphi_{p-1}(\lambda)| < K^{p-1} M \frac{\lambda^p}{p!}.$$

Thus, the series of $\sum_p (\varphi_p(\lambda) - \varphi_{p-1}(\lambda))$ converges, which implies the existence of the $\lim_{p\to\infty} \varphi_p(\lambda)$ uniformly in $\lambda \in [0, \epsilon]$.

Coming back to (7.5.4), we finally come to (7.5.3) by taking the right derivative in the variable λ at $\lambda = 0$ and by noting $\varphi_0(\lambda)$ is constant.

Note that the computation for the proof are uniformly in η, that is independent of the direction of change of ξ. Hence the existence has been proved.

As for the uniqueness we can prove by using the assumption (ii) of (A.2).

So far, we have discussed classical theory of variations for non-random fields. To apply the results to stochastic variational calculus, we have to apply transformations S and S^{-1} before and after the theory discussed in this section.

If a random field is given in the form $X(\xi)$, then applying the S-transform to have $U(\xi)$, so that we are ready to come to the variational calculus discussed above. If the given random field is of the form $X(C)$, then we let C be represented by a vector-valued function. Slight generalization is possible by using the results obtained so far.

When the solution of the (non-stochastic) variational equation is obtained, it is necessary to check the Potthoff–Streit criterion (see Section 2.1) to find if the solution is a U-functional.

Finally, we have to check that our theory is independent of the choice of a vector-valued function. In fact, this can be done easily.

Summing up those steps, we come to the complete theory of stochastic variational equations.

7.6 A generalization of the Ito formula for Gaussian random fields

We shall be concerned with a random field which is a function of a Gaussian Markov field or with a field of homogeneous chaos, which is Markov in the sense of Section 6.2.

Now take a Gaussian Markov field $X(C)$. By Theorem 5.8, it is expressed in the form $X(C) = f(C)Y(C)$, where $f(C)$ is non-random and $Y(C)$ is a martingale. So we are concerned with the martingale $Y(C)$ to establish Ito formula.

Let $Y(C)$ be a Gaussian martingale with respect to the σ-fields $\mathbf{B}_C(Y)$ such that $Y(C) \neq 0$ for every C. Then, as was seen before, the $Y(C)$ is expressed in the form, with a suitable choice of $g(u)$,

$$Y(C) = \int_{(C)} g(u)x(u)du^d,$$

and its variance is

$$\sigma^2 = \sigma((C)) = \int_{(C)} g(u)^2 du^d.$$

Define

$$Z(C) = H_n(Y(C); \sigma^2),$$

where $H_n(x; \sigma^2)$ is the Hermite polynomial with parameter σ^2.

Proposition 7.3 *The variational equation for $Z(C)$ is given by*

$$\delta Z(C) = H_{n-1}(Y(C); \sigma^2) \int_C g(s)x(s)\delta n(s)ds.$$

Proof. Take the S-transform of $Z(C)$, then we have the U-functional

$$U(C, \xi) = \frac{1}{n!} \int_{(C)^n} g^{n\otimes}(u)\xi^{\otimes n}(u)du^n,$$

where $u = (u_1, u_2, \ldots, u_n)$. Its variation is

$$\delta U(C, \xi) = \frac{1}{(n-1)!} \int_{(C)^{n-1}} g^{(n-1)\otimes}(u')\xi^{\otimes(n-1)}(u')du^{n-1} \times \int_C g(s)\xi(s)\delta n(s)ds,$$

where $u' = (u_1, u_2, \ldots, u_{n-1})$. Applying the S^{-1}-transform, the required result is obtained.

Remark 7.2 *Such a simpler formula is obtained because of the favourable property of Hermite polynomial.*

We now consider a variation of a smooth (L^2)-functional $F(X(C))$ of a Gaussian Markov random field $X(C)$. Since $X(C)$ is Markov, it is expressed in the form $X(C) = f(C)Y(C)$ was noted before. Hence there exists a $\mathbf{B}(X)$-measurable (L^2)-functional G such that

$$F(X(C)) = G(C, Y(C)).$$

Since it is a (nonlinear) function of a Gaussian variable $Y(C)$, we have an expansion in terms of the Hermite polynomials in $Y(C)$:

$$G(C, Y(C)) = \sum a_k(C) H_k(Y(C); \sigma^2).$$

Theorem 7.4 *Assume that*

$$\sum \left[\left(\frac{\partial a_k(C)}{\partial n}(s) \delta n(s) ds \right)^2 + a_{k+1}^2 \right] \frac{\sigma_k^2}{k!}$$

converges. Then we have

$$\delta F(X(C)) = \sum \left(\int_C \frac{\partial a_k}{\partial n}(s) \delta n(s) ds \right) H_k(Y(C); \sigma^2) \\ + \sum a_k(C) H_{k-1}(Y(C); \sigma^2) \int_C f(s) x(s) \delta n(s) ds.$$

The proof is given in a usual manner.

7.7 The Poisson case

The Poisson noise is another basic elemental generalized stochastic process with independent values at every t. The characteristic functional $C_P(\xi)$ of Poisson noise $\dot{P}(t)$ is given by (3.1.1) and it is modified to the centered version expressed by (3.1.2).

The Hilbert space formed by complex valued functionals of $\dot{P}(t)$'s with finite variance will be denoted by $(L^2)_P$. The analysis on this Hilbert space can be proceeded in a similar manner to the Gaussian case. So, we do not go into details, since the similarity is not so interesting.

We then come to a multi-parameter (say, d-dimensional parameter) Poisson noise, denoted by $V(u), u \in R^d$. The characteristic functional is of the same form in expression as in the one dimensional parameter case.

$$C_P(\xi) = \exp \left[\lambda \int_{R^d} (e^{i\xi(t)} - 1) dt \right], \quad \xi \in E,$$

from which one can see a sample function $x(u)$ (the notation $\dot{P}(u)$ is not fitting in the R^d-parameter case) involves randomly arranged delta functions only. The $C_P(\xi)$ determines a probability measure ν_P on E^*.

The innovation problem for a random field $X(C)$:

$$X(C) = \int_{(C)} f(C, u)x(u)du^n$$

admits the variation

$$\delta X(C) = \int_C f(C, s)x(s)\delta n(s)ds + \int_{(C)\times C} \delta f(C, u)(s)x(u)\delta n(s)du^d ds.$$

Now, the innovation $x(s), s \in C$, of $X(C)$ can be obtained in the similar manner to the Gaussian case, having assumed that the integral defining the $X(C)$ is a stochastic integral with respect to the random measure $x(u)du^d$.

On the other hand, as a generalization of the definition of the integral to be a continuous linear functional explained in Section 6.3, the path-wise integral can be defined, although it is still expressed in the same notation, if the kernel function $f(C, u)$ is a smooth function of u for every C.

One may wonder if the following observation has any contradiction. That is, the sample function of the innovation $x(s), s \in C$, is a random sequence of delta-functions sitting on C, whilst it is to be obtained by restricting the parameter $u \in R^d$ of the original innovation to the manifold C, and it is also a random sequence of delta-functions. Note that the original Poisson noise is not simply a sum of random delta-functions on C. The reader reminds the discussion given in Section 2.6.

Interesting observation of Poisson noise with R^d-parameter is given by taking a Poisson sheet. It is the Poisson functional $X(t) = \langle x, \chi_{I(t)} \rangle = \langle V, \chi_{I(t)} \rangle$ is defined as a stochastic bilinear form like in Gaussian case, where $I(t) = \prod_{j=1}^d [0, t_j], t = (t_1, \ldots, t_d)$; and where $x \in E^*(\nu_p)$.

Suppose $X(\mathbf{1}) = n, \mathbf{1} = (1, \ldots, 1)$. Then, the jump points of $X(t)$ in the interval correspond to the δ-functions of $V(t)$. The positions where they are sitting are random. Note that between the two nearest neighbour delta functions in some coordinate direction, there is an exponential distribution. However, if we assume that only n delta-functions are there and they are distributed randomly, then we claim

Proposition 7.4 *The distribution of any fixed coordinate of n random delta-functions is uniform on the simplex $u_1 + u_2 + \cdots + u_d = n, u_i \geq 0$,* *which is compact and is a part of the hyper-plane in R^d. Those d coordinates are mutually independent.*

Proof. The partial derivative of the Poisson sheet in one of the variables gives a delta-function and we have such delta-functions as many as n, under the restriction that the sum of their intervals is equal to 1. Then, it is easy to see that they are uniformly distributed over the n-dimensional simplex. Then, such an observation is made for each derivation to come to the conclusion. The rest of the assertion is obvious.

By using $X(t)$, we can easily restrict the parameter to a lower dimensional set.

7.8 Characteristic functionals

We now discuss characteristic functionals of random fields with multi-dimensional parameter. The space where general random fields in question are living is the space $(\mathbf{P})$ for which existence of variance is not assumed. In such a case, we often use the characteristic functional which includes all the information on the probability distributions of the random field. Indeed, the characteristic functional will never face fundamental limits to play their roles.

The information source of those random fields to be discussed is generally assumed to be white noise (that is Gaussian, Poisson, compound Poisson or mixture). We then assume that if a sample function is observed its characteristic functional can be obtained theoretically from the observed sample function (path).

There are several significant cases where the characterization of the random field is given by the functional properties of its characteristic functionals. They can be seen in the following.

(1) White noise

Assume that the associated measure is (infinite dimensional) rotation invariant and is ergodic, then the characteristic functional should be of the form

$$C(\xi) = \exp\left[-\frac{\sigma^2}{2}\|\xi\|^2\right],$$

that is a Gaussian measure or a trivial measure ($\sigma^2 = 0$). (See, e.g. [18, Section 5.6].)

(2) Poisson case

The characteristic functional is, as we have seen before, of the form

$$C_P(\xi) = \exp\left[\lambda \int_{R^d} (e^{i\xi(t)} - 1)dt\right].$$

A characterization in the case $d = 1$ was given in Section 3.2 in terms of a functional equation.

Some invariance of the distribution of delta functions which are consisted in the sample function of Poisson noise has also been discovered.

(3) The Hopf equation (See [39].)

Consider a liquid running on a region R with boundary ∂R. The velocity field is defined by

$$u^t = u^t(x) = u(x,t), \quad u^t \in R^3, \;\; x \in R + \partial R, \;\; t \geq 0,$$

$$u^0 = u, \quad T^t u = u^t,$$

where $\{T_t, t \in R\}$ is a *flow* on the velocity field Ω. The field is solenoidal, that is

$$\operatorname{div} u = 0.$$

The equation that governs the flow of the liquid is

$$u_{j,k} + u_k u_{j,k} = -p_j + \mu u_{j,k}.$$

Let P be the phase distribution on Ω and define P^t by

$$P^t(B) = P(T_{-t}B), \quad B \subset \Omega.$$

The $\{P^t\}$ determines the dynamical system. The characteristic functional Φ of this phase distribution is given by

$$\Phi(\xi, t) = \int_\Omega \exp[i\langle \xi, u\rangle] P^t(du).$$

Then we have

$$\frac{\partial \Phi}{\partial t} = \int_R \xi_j(x)\left[i\frac{\partial}{\partial x_k}\frac{\partial^2 \Phi}{\partial \xi_k(x)dx\partial \xi_j(x)dx} + \mu\Delta_x \frac{\partial \Phi}{\partial \xi_j(x)dx} - \frac{\partial \Pi}{\partial x_j}\right] dx,$$

where Π is determined by the boundary conditions and others and where we apply the summation rule in the pair of the same indices.

To solve this equation, there have been many attempts and met difficulty when the Φ is expanded into Taylor series in terms of monomials in ξ. There appears a singularity, which we have met in the case of generalized white noise functionals. There is a hope that the same idea to manage such singularity is applicable.

(4) A stochastic process of mixed type

$$X(t) = \varphi(t, \dot{B}) + \psi(t, \dot{P}).$$

The analytic expression of the characteristic functional can be factorized into the two components. Gaussian part corresponds to a polynomial in ξ, while Poisson part involves the functional $e^{i\xi}$.

To discuss the innovation of $X(t)$ the martingale theory (Meyer decomposition) can effectively be used. (See [2].)

If φ and ψ are linear, that is, if $X(t)$ is a linear process, we will discuss in more details in Section 8.4. Further, it is known that the characteristic functional of $X(t)$ determines its probabilistic structure. (See [94] and the forth coming paper by the same author.)

(5) Random field $X(C)$

Let $X(C) = X(C, x)$, $C \in \mathbf{C}$, be a random field, where x is a R^d-parameter white noise.

Define a characteristic function

$$\varphi_{(C_1,\dots,C_n)}(z_1, \dots, z_n) = E\left[e^{i\sum z_j X(C_j)}\right] \tag{7.8.1}$$

where $z = (z_1, \dots, z_n) \in R^n$. Then we have a *consistent* family of probability measures on $(R^{\mathbf{C}}, \mathbf{B}^{\mathbf{C}})$, where $\mathbf{B}^{\mathbf{C}}$ is the σ-field generated by cylinder subsets of $R^{\mathbf{C}}$. We can apply the Kolomogorov extension theorem which guarantees the existence of probability measure ν on $(R^{\mathbf{C}}, \mathbf{B}^{\mathbf{C}})$.

Let $T_t^{(j)}, j = 1, 2, \dots, d$, be the flows defined by the shifts S_t^j of R^d, under which white noise is invariant.

The $X(C)$ is said to be the T_t^j stationary if $\{X(C, T_t^j x),\ C \in \mathbf{C}\}$ and $\{X(S_t^j C, x)\}$ have the same probability distribution. Namely,

$$T_t^j \nu = \nu, \quad \text{for every } t. \tag{7.8.2}$$

Let $Z(C)$ be a real-valued function and let m be a σ-finite measure on $\mathbf{C}$. Assume that

$$\int Z(C) X(C) dm(C)$$

is defined and is continuous in C. Then we define the characteristic functional $C_X(Z)$ of $X(C)$.

We are now going to establish a generalization of the Bochner–Minlos theorem to show a one to one correspondence

$$C_X(Z) \leftrightarrow \nu.$$

To this end we prepare some background.

Let $E = \{Z(C),\ C \in \mathbf{C}\}$ be a collection of $Z(C)$ which is differentiable in C infinitely many times. Here, derivative means the Fréchet derivative. $Z'(C,s)$ is the the derivative if $\delta Z(C)$ is expressed in the form

$$\delta Z = \int_C Z'(C,s)\delta n(s)ds. \tag{7.8.3}$$

We assume that $Z'(C,s)$ is continuous in (C,s) and

$$\int\int_{\mathbf{C}} |Z'(C,s)|^2 dm(C)ds < \infty. \tag{7.8.4}$$

In the same manner, the higher order derivative are defined, and we claim

$$\|Z(C)\|_n^2 = \int \cdots \int_{C\times C^n} |Z^{(n)}(C,s_1,\ldots,s_n)|^2 dm(C)ds_1\cdots ds_n < \infty.$$

The seminorms $\|\ \|_n, n \geq 0$, define the topology for E, and it is proved that E with this topology is a nuclear space.

Theorem 7.5 *The functional $C_X(Z), Z \in E$, is positive definite. Hence, there exists a probability measure ν such that*

$$C_X(Z) = \int e^{i\langle Z,x\rangle} d\nu(x), \tag{7.8.5}$$

where x stands for a sample function of a random field parameterized by $C \in \mathbf{C}$.

Chapter 8

Innovation Approach

8.1 Concept of innovation

This chapter continues to occupy the central part of what we are going to explain in this book. We shall start with some motivation on the innovation approach.

The original intention of our white noise theory is the investigation of evolutional random complex systems under a general set up. Typical examples of such a system are stochastic processes $X(t), t \in R$, random fields $X(C)$ parameterized by a contour or a surface C and their generalizations. We are particularly interested in random fields $X(C)$, C being a contour. The basic idea of the analysis of the system is the use of the *innovation.* With this idea functionals of general innovation will appear in order to express the given system $X(C)$ in terms of the innovation, and the general theory of the functional analysis can analyze those functionals. Thus, we are naturally requested to construct the innovations of the given random complex systems.

Now let us first recall again the notion of innovation for a stochastic process $X(t)$. P. Lévy introduced in 1953 (see [48]) the so called *stochastic infinitesimal equation* expressed in the form

$$\delta X(t) = \Phi(X(s), s \leq t, Y(t), t, dt), \tag{8.1.1}$$

as was illustrated in Chapter 1. See also P. Lévy [45] for the discrete parameter case.

It is requested that the $Y(t)$ (scalar or vector valued idealized random variable) contains as much information as the $X(t)$ gains during the infinitesimal time interval $[t, t + dt)$. So far we have considered the notion of the innovation in an intuitive level. A mathematically rigorous definition is now given below. It would also be mathematically satisfactory.

Definition 8.1 The system $\{Y(t)\}$ is the innovation of a stochastic process $\{X(t)\}$ if the following conditions are satisfied.

1. It is a generalized stochastic process with independent values at every moment,
2. $Y(\xi) = \langle Y, \xi \rangle$ is measurable with respect to $\mathbf{B}_{t+}(X)$ where t is the supremum of the supp(ξ), the support of ξ, and independent of $\mathbf{B}_{s-}(X)$, where s is the infimum of the supp(ξ),
3. It holds that

$$\mathbf{B}_{t-}(X) \bigvee \left(\bigwedge_{t \in \mathrm{supp}(\xi)} \mathbf{B}(Y(\xi)) \right) = \bigwedge_{\epsilon > 0} \mathbf{B}_{t+\epsilon}(X).$$

Remark 8.1 *The $Y(t)$ can be a random vector (finite or infinite dimensional) or a linear combination of different generalized processes provided that components can locally be separated by probabilistic method, as we can see in what follows. The situation is the same in the case of random fields.*

Once the innovation is obtained we expect that the given process $X(t)$ could be expressed as a function of the $Y(t)$'s. It should be emphasized that the $Y(t)$'s can be taken to be the variables of the function that represents the given phenomenon. If this is realized, we shall be ready to analyze the given stochastic process.

A possible generalization of the stochastic infinitesimal equation for a random field $X(C)$ depending on a contour (or a loop) C may be, as was briefly explained in Chapter 1, proposed to be an equation expressed in the form

$$\delta X(C) = \Phi(X(C'), C' < C, Y(s), s \in C, C, \delta C), \tag{8.1.2}$$

where $C' < C$ means that C' is inside of C, that is, the domain (C') enclosed by a contour C' is a subset of (C), and where Φ is, as before, a non-random function, and the system

$$Y = \{Y(s), s \in C; C \in \mathbf{C}\}$$

is the *innovation* of $X(C)$.

To be more clear, the definition is given as follows. First, the notations are to be fixed.

Let $\mathbf{B}_C(X)$ be the sigma-field generated by all the $X(C')$ with $C' < C$. Similarly, $\mathbf{B}_C(Y)$ is defined in such a way that

$$\mathbf{B}_C(Y) = \bigwedge_{\text{supp}(\xi) \supset (C)} \mathbf{B}(Y(\xi)).$$

Definition 8.2 The *innovation* of $X(C)$ is a family of systems $Y(C) = \{Y(s), s \in C\}$, with $C \in \mathbf{C}$, such that

(1) $Y(s)$ is a generalized stochastic process parameterized by R^d and has independent values at every point s.
(2) $Y(\xi), \xi \in E$, is independent of $\mathbf{B}_C(X)$ if $\text{supp}(\xi) \cap \overline{(C)} = \phi$, and $Y(\xi)$ is $\mathbf{B}_C(X)$-measurable if $\text{supp}(\xi) \subset (C)$.
(3) It holds that

$$\mathbf{B}_{C-}(X) \vee \left(\bigwedge_{\xi} \mathbf{B}(Y(\xi)) \right) = \bigwedge_{\delta n > 0} \mathbf{B}_{C+\delta C}(X),$$

where ξ satisfies $\text{supp}(\xi) \supset C$.

We understand the important deformation δC of C is represented by a system

$$\delta C \simeq \{\delta n(s), s \in C\},$$

where $\delta n(s)$ denotes the *length* of normal vector to C at s.

For either $X(t)$ or $X(C)$ we can prove the following:

Triviality. The sigma-fields generated by the innovation is unique if innovation exists.

There are, of course, many choices of a system of **elemental random variables** $Y(s)$, but we claim the uniqueness in terms of the σ-field.

We also consider the case where X depends on a higher dimensional vector or on a function f defined on an interval $[a, b]$ such that $f(a)$ and $f(b)$ are fixed. The former has been discussed and the theory has been established. The second case, we can employ the same technique as in the non-random case. Note that the parameter f has fixed boundary value, so that we can avoid the awkwardness arising from the singularity at the boundaries.

Note that the equation (8.1.2) has, as in the case of the stochastic infinitesimal equation, only a formal significance. In order to give correct interpretation or understanding, we must specify the class of fields, that is the class of $X(C)$'s for which the innovation actually exists. Before doing so, we shall first deal with a simple case in order to give a plausible interpretation to the idea of our innovation approach to random fields. We also recall the aims of the variational calculus stated in Chapter 7; one of them is to form the innovation.

8.2 Lévy decomposition of innovation

Innovation has a basic probabilistic property. The case of a stochastic process and that of a random field are studied separately, although the idea is the same.

(1) One dimensional parameter case; i.e. a stochastic process
The innovation of a stochastic process $X(t), t \in R$, is a generalized stochastic process with independent values at every t. This fact is shown by definition. It is reasonable to assume that it is the time derivative of a Lévy process, denoted by $L(t), t \in R$. In other words, integration of innovation is assumed to give us a Lévy process.

$$\textbf{Innovation} \quad \rightarrow \quad \textbf{Lévy process } L(t).$$

It is well known that a Lévy process, under mild assumptions, admits a decomposition of the form

$$L(t) = m(t) + X_0(t) + X_1(t),$$

where $m(t)$ is a non-random function, $X_0(t)$ is an additive Gaussian process and $X_1(t)$ is a compound Poisson process. In addition, those processes are mutually independent. One may ignore the sure function $m(t)$. If $X_0(t)$ and $X_1(t)$ are assumed to have stationary (independent) increments with mean 0, then $X_0(t)$ is a Brownian motion up to constant, say $cB(t)$, $c > 0$, and the $X_1(t)$ admits further decomposition into independent Poisson processes $P_u(t)$ with different heights of jumps. It is noted that $X_1(t)$ is a superposition of the $P_u(t)$'s, where the sum is defined as a quasi-convergence in probability. For details of this fact we refer to P. Lévy [46] Chapter V.

From our standpoint *reduntionism*, the components $B(t)$, $P_u(t)$'s of a Lévy process are all viewed as **elemental processes**, in fact, elemental generalized stochastic processes are obtained by taking their derivatives.

When the functionals of the innovations are discussed, we can start with functionals of each elemental process separately, then the results are combined. This is the way we shall do.

Remark 8.2 *This is a short note which would have connection in the future study. We are interested in not only the analysis on (L^2), but also sample function-wise calculus. In the latter case, it is necessary to obtain sample functions of each $P_u(t)$, namely to obtain the instants when $P_u(t)$ jumps. This problem is often called "jump finding problem". The results will be used in the discussion of computability.*

(2) Multiple-parameter case

(i) For a random field $X(a), a \in R^d, (d > 1)$, the radial direction is considered as the address of evolution. Unless $X(a)$ is degenerated, it has countably many multiplicity regarding the evolution. Associated with each cyclic subspace with unit multiplicity is an innovation. Theoretically speaking, the same game as in (1) will be played as many time as infinite. A good example is the Lévy's Brownian motion, for which one can see in the McKean's result [56].

(ii) For a random field $X(C)$ parameterized by C, an interpretation has been given in the previous section with the help of the stochastic variational equation. We are now ready to discuss the innovation in this line.

8.3 Review of linear parameter case

To discuss the topic of this section, it is necessary to review some notes on the integral based on a Brownian motion $B(t)$. We shall use two different kind of integrals based on $B(t)$.

(i) $dB(t)$, considered as a random measure

To let the present article be self contained, here is given a brief interpretation on stochastic integral.

The integral is defined like the ordinary integral over a time-interval T. For an interval $\Delta = (a, b]$, the increment of $B(t)$ over Δ is denoted by ΔB, which is $B(b) - B(a)$. For a simple function $f(u) = \sum k_j \chi_{\Delta_j}(u)$ we define an integral

$$I(f) = \sum k_j \Delta_j B.$$

The $I(f)$ is denoted by $\int f(u)dB(u)$ in an integral form. The integral is a Gaussian random variable with mean 0 and variance $|f|^2$ as is early seen. If f_n strongly converges to f in $L^2(T)$, then $I(f_n)$ forms a Cauchy sequence in $L^2(\Omega, P)$. Hence the strong limit of $I(f_n)$ exists. This limit is independent of the choice of f_n and is denoted by $I(f)$ or by

$$\int f(u)dB(u),$$

and is called the *Wiener integral* of f. The integral is a particular case of a stochastic integral having a non-random integrand. The $dB(u)$ is therefore well defined and is a most important example of a random measure. By many reasons we prefer the notation $\dot{B}(u)du$ rather than $dB(u)$.

(ii) Stochastic bilinear form

A sample function of $B(t)$ is continuous but not differentiable. Hence, the time derivative $\frac{d}{dt}B(t) = \dot{B}(t)$ has meaning only as a generalized function. Hence, a test function ξ in a space E involving sufficiently smooth functions, say a nuclear space, enables us to define an ordinary random variable, namely, a canonical bilinear form $\langle \dot{B}, \xi \rangle$ gives a rigorous meaning. It is often denoted in the form

$$\int \xi(t)\dot{B}(t)dt.$$

If $x(\in E^*)$ is a sample function of $\dot{B}$, then we use a notation

$$\int \xi(t)x(t)dt.$$

If ξ_n tends to f in $L^2(T)$, then $\int \xi_n(t)x(t)dt$ converges to some random variable, which is denoted by $\int f(t)x(t)dt$. This is called a *stochastic bilinear form*. Of course, this is in agreement with the Wiener integral defined in (i) if it is viewed as a member of (L^2).

Let $\{X(t)\}$ be a Gaussian process with one dimensional parameter $t \in T \subset R^1$ satisfying the condition $E(X(t)) = 0$. Assume, in particular, that the $X(t)$ has a representation in terms of a white noise $\dot{B}(t)$ as a Wiener integral of the form

$$X(t) = \int^t F(t,u)\dot{B}(u)du, \quad t \in T, \tag{8.3.1}$$

where the kernel $F(t,u)$ is square integrable in u for every t, and is smooth in t. Then, its variation over an infinitesimal time interval $[t, t+dt)$ is given by

$$\delta X(t) = F(t,t)\dot{B}(t)dt + dt\int^t F_t(t,u)\dot{B}(u)du + o(dt), \tag{8.3.2}$$

where $F_t(t,u) = \frac{\partial}{\partial t}F(t,u)$.

As is well known, a representation of the form (8.3.1) is not unique for a given $X(t)$. Let us take the *canonical* representation, which gives some advantage to our innovation approach. With such a choice of the representation, it satisfies the condition

$$E\big[X(t)|\mathbf{B}_s(X)\big] = \int^s F(t,u)\dot{B}(u)du, \quad \text{for any } s < t, \tag{8.3.3}$$

where $\mathbf{B}_s(X)$ is the smallest σ-field with respect to which all the $X(u)$, $u \leq s$, are measurable.

Note that a generalization of the representation theory to the Gaussian random fields has been used in Chapter 5.

Proposition 8.1 *If a Gaussian process has a representation of the form* (8.3.1), *the function* $F(t,t)^2$ *is uniquely determined regardless the representation is canonical or not.*

Proof. The variance of $X(t+dt) - X(t)$ is $F(t,t)^2 dt + o(dt)$, which is independent of the way of representation. Hence, the assertion is proved.

We have a freedom to choose the sign of $F(t,t)$, but we do not care the sign, since $\dot{B}(t)dt$, which is to be associated to $\sqrt{dt}$, has symmetric probability distribution.

Assume that

$$\delta X(t) \text{ is of order } \sqrt{dt}. \tag{8.3.4}$$

This means that $X(t)$ is not differentiable since $F(t,t) \neq 0$ and since $\delta X(t)$ has non-trivial randomness. Then, the first term of (8.3.2) is non-vanishing. Hence $F(t,t)$ is not zero and it may be taken to be positive and continuous. With this assumption and with the note that $X(t)$ has unit multiplicity (which is equivalent to the existence of the canonical representation), we can prove the following theorem.

Theorem 8.1 *For a representation of* $X(t)$, *assume that the boundary value of the kernel* $F(t,t)$ *never vanishes and is continuous. Then the limit*

$$\lim_{dt \to 0+} \frac{\delta X(t) - E[\delta X(t)|\mathbf{B}_t(X)]}{F(t,t)} \tag{8.3.5}$$

gives the innovation.

Proof. First we introduce the stochastic derivative of $X(t)$ (For definition, see [2].). It is meaningful in the case where $X(t)$ is not differentiable, so that non-differentiability of $X(t)$ is assumed. Let

$$D_+X(t) = \lim_{\epsilon \to 0+} \frac{E(X(t+\epsilon)|\mathbf{B}_t(X)) - X(t)}{\epsilon}, \quad \text{a.e}$$

be the stochastic derivative of $X(t)$. Then, we have

$$X(t) = \int_0^t D_+X(u)du + M_t,$$

where M_t is a martingale. In our case it is actually an additive Gaussian process expressed in the form

$$M_t = \int_0^t F(u,u)\dot{B}(u)du.$$

Taking the time derivative, we get $\dot{B}(t)$, which is to be the innovation. In short, we may write in the form (8.3.5). (Also, see [2].)

Remark 8.3 *The random element in the numerator in* (8.3.5) *is obtained by subtracting the past value from the infinitesimal increment of the* $X(t)$, *hence it stands for the new random variable appeared at instant* t. *It is therefore the innovation. The innovation thus obtained will be denoted by the same symbol* $\dot{B}(t)$ *as was used in* (8.3.1). *However, we should note that it may be different from the original white noise, if the representation* (8.3.1) *is not the canonical representation.*

Once the $\dot{B}(t)$ is given for every t, we can use the differential operator given by (2.1.15):

$$\partial_u = \frac{\partial}{\partial \dot{B}(u)}, \quad u \le t. \tag{8.3.6}$$

Apply ∂_u to $X(t)$ to have $F(t,u)$:

$$\partial_u X(t) = F(t,u), \quad u \le t.$$

This $F(t,u)$ is the canonical kernel that we are looking for. Noting that $\dot{B}(t)$ is the innovation, we can establish the following proposition.

Proposition 8.2 *The exact value of the canonical kernel* $F(t,u)$ *is obtained by applying the operator* ∂_u, $u \le t$, *to the* $X(t)$.

Thus we can see that the expression (5.1.1) for the canonical representation can be completely determined through the determination of the innovation, and hence the structure of the given Gaussian process $X(t)$ can be known.

We now come to a *non-canonical* representation which is given by a stochastic integral

$$X(t) = \int_0^t G(t,u)\dot{B}(u)du, \tag{8.3.7}$$

where the kernel $G(t,u)$ is a non-canonical kernel and is assumed to be continuous in (t,u) and smooth in u for every t. Consider a particular case where $G(t,t)$ never vanishes. This condition is always assumed in what follows.

In fact, this assumption does not depend on either the representation is canonical or not. Here the integral can be regarded not as a Wiener integral, but as an integral defined path-wise. Namely it is a continuous bilinear form of $G(t,\cdot)$ and $\dot{B}(\cdot)$.

As before, the variation of $X(t)$, given by (8.3.7) is anyhow given by

$$\delta X(t) = G(t,t)\dot{B}(t)dt + dt\int_0^t G_t(t,u)\dot{B}(u)du, \tag{8.3.8}$$

with $G_t(t,u) = \frac{\partial}{\partial t}G(t,u)$. Taking the quadratic variation of $\delta X(t)$ we obtain $G(t,t)^2$. Take the square root to have the exact values of $|G(t,t)|$. Then, we have

$$\frac{\delta X(t)}{|G(t,t)|dt} \to \dot{B}_1(t)$$

as $dt \to 0$. The limit $\dot{B}_1(t)$ is equal to the original $\dot{B}(t)$ up to sign. We may therefore call it the *quasi-innovation* and may not be the real innovation. It is noted that to get the *quasi-innovation* the nonlinear operation (i.e. quadratic variation) is requested.

Proposition 8.3 *From the non-canonical representation defined by the path-wise integral, the quasi innovation is obtained, provided that the kernel never vanishes at* $t = u$.

Remark 8.4 *It is known that the information of white noise is partly lost through the filter* G*. The lost information cannot be recovered causally, however we can form an equivalent white noise to subtract off some information. We may also think of the equivalence of the sigma-fields generated by the process and the quasi innovation.*

8.4 Innovations of linear processes

A definition of a process with linear correlation was given by P. Lévy [51] in 1957, however we wish to discuss somewhat restricted class and call a process in this class simply a linear process. Details of the definition is not given here, since we wish to go directly into the question on innovation.

We take stochastic processes which are expressed as a sum of linear functions of Gaussian and Poisson noise. More precisely, let $Z(t)$ be a linear process given by

$$Z(t) = X_1(t) + X_2(t), \quad X_1(t) \in \mathbf{X}_1, \; X_2(t) \in \mathbf{X}_2, \tag{8.4.1}$$

where $\mathbf{X}_1$ is a collection of Gaussian processes with unit multiplicity (that is, with the canonical representation) and

$$\mathbf{X}_2 = \left\{ X_2(t) = \int_0^t G(t,u)\dot{P}(u)du, \quad G \in \mathbf{C}^2, \; G(t,t) \neq 0 \right\}. \tag{8.4.2}$$

The analysis of such a linear process $Z(t)$ is discussed in the space $(\mathbf{P})$ which is now taken to be collection of $\mathbf{B}(\dot{B}, \dot{P})$-measurable random variables. It is topologized, so has been determined before, by the convergence in probability. Certainly, the present analysis is different from that is done on Hilbert space like (L^2) defined by a Gaussian process.

Proposition 8.4 *The expression of $X_2(t)$ in $\mathbf{X}_2$ is always a canonical representation in the sense that*

$$\mathbf{B}_t(X_2) = \mathbf{B}_t(\dot{P})$$

holds for every t.

Proof. Each discontinuous point of $X_2(t)$ is a jump point of $P(t)$, and vice versa. Since $P(t)$ is determined only by jump points and since $X_2(t)$ is a functional of $P(t)$, we conclude the equality for sigma-fields.

Definition 8.3 A stochastic process $Z(t)$ which is expressed as a sum of the form

$$Z(t) = X(t) + Y(t), \quad t \geq 0, \; X(t) \in \mathbf{X}, \; Y(t) \in \mathbf{Y},$$

is called a *linear process.*

Theorem 8.2 *Let a linear process $Z(t)$ be given. Then,*

(1) *the two terms of a linear process $Z(t)$ can be formed by $Z(s), s \leq t$, and*
(2) *$a\dot{B}(t) + b\dot{P}(t)$ is an innovation, where a and b are any non-zero constants.*

Proof. The two assertion can be proved by observing the sample function of the process $Z(t)$.

Remark 8.5 *For details see Win Win Htay [94].*

Remark 8.6 *Proof of Theorem 8.2 may be given in a similar manner to the proof of Theorem 8.1, by using the stochastic derivative and Proposition 8.4. Actually, the martingale M_t in this case is given by*

$$M_t = \int_0^t F(u,u)\dot{B}(u)du + \int_0^t G(u,u)\dot{P}(u)du,$$

from which the innovation is obtained.

Once the innovation is obtained, it is equivalent to have both a $\dot{B}(t)$ and $\dot{P}(t)$ (like the case of the Lévy decomposition of a Lévy process). The component processes $X_1(t)$ and $X_2(t)$ of $Z(t)$ have now been at hand. Hence, for $s \leq t$, the canonical kernels $F(t,u)$ and $G(t,u)$ can be computed theoretically by:

$$\frac{\partial}{\partial \dot{B}(u)} X_1(t) (= \partial_u X_1(t)) = F(t,u),$$
$$\frac{\partial}{\partial \dot{P}(u)} X_2(t) (= \partial_u X_2(t)) = G(t,u).$$

The second formula implies

Corollary 8.1 *The (kernel of the) representation of $X_2(t)$ is unique.*

Remark 8.7 *Uniqueness of this representation may be proved by conditional expectation.*

Example 8.1 A linear process formed by a Poisson noise $\dot{P}(t)$. Set

$$X(t) = \int_0^t G(t,u)\dot{P}(u)du,$$

where $G(t,u)$ is smooth on the domain $D_t = \{(t,u), 0 \leq u \leq t\}$ and $G(t,t) \neq 0$ for $t > 0$. The integral is understood to be a bilinear form $G(t,\cdot)$ and $\dot{P}(\cdot)$ and is defined sample function-wise in the sense of ii) in

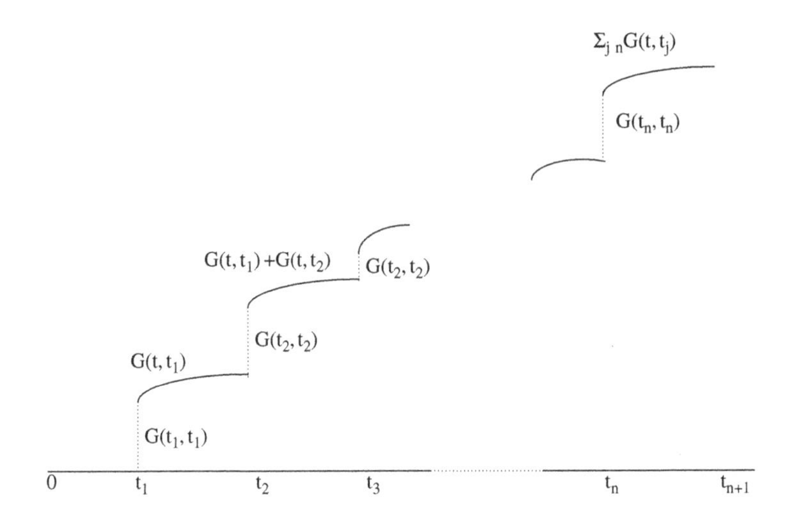

Fig. 8. Sample function of $X(t)$.

Section 8.3. For a fixed random parameter denoted by ω, $\dot{P}(t,\omega)$ is a sum of delta-functions, i.e.

$$\dot{P}(t,\omega) = \delta_{t_j}(t), \quad t_j = t_j(\omega).$$

The t_j's correspond to the times when $\dot{P}(t,\omega)$ jumps.

Hence we have

$$X(t) = \sum_{t_j \le t} G(t, t_j).$$

$X(t)$ jumps at t_j with the height $G(t_i, t_j)$, which is non-zero by assumption.

Here we should note that the characteristic function $C_X(\xi)$ of $X(t)$ is of the form

$$C_X(\xi) = \exp\left[\lambda \int (\exp(i(\check{G} * \xi)(u)) - 1)du\right], \tag{8.4.3}$$

where

$$(\check{G} * \xi)(u) = \int_u^\infty G(t,u)\xi(t)dt.$$

By Proposition 8.4, we can see that the characteristic functional determines the kernel $G(t,u)$ uniquely.

8.5 Innovation of a linear random field

Consider a linear random field $Z(C), C \in \mathbf{C}$, which is expressed as a sum

$$Z(C) = X_1(C) + X_2(C), \quad X_1 \in \mathbf{X}_1, \ X_2 \in \mathbf{X}_2, \tag{8.5.1}$$

where $\mathbf{X}_1$ involves Gaussian random field $X_1(C)$ which has the canonical representation

$$X_1(C) = \int_{(C)} F(C,u)W(u)du^d,$$

and where $\mathbf{X}_2$ is a collection of $X_2(C)$ such that

$$X_2(C) = \int_{(C)} G(C,u)V(u)du^d,$$

expressed in terms of a Poisson noise V. The $\mathbf{C}$ is taken to be a collection of smooth ovaloid in R^d, as before. Then, we can prove, by observing sample functionals.

Proposition 8.5 *Given the value of $Z(C)$. Then*

(i) *the two components in* (8.5.1) *of $Z(C)$ can be constructed from $Z(C)$.*
(ii) *Innovation is given to be the pair of the systems $W(s), s \in C$ and $V(s), s \in C$.*

8.6 Stochastic variational equations

We then come to the main subject of this chapter, that is the variational calculus for random fields. Here we remind the notations. Let $X(C)$ be a random field with parameter C which runs through the class $\mathbf{C}$, as was prescribed before. White noise with parameter space R^d is denoted by $x(u), u \in R^d, x \in E^*$. Here, E^* is the space of generalized functions on R^d and the white noise measure μ is introduced on E^*. The $X(C)$ in question is a functional of x.

We are interested in the variation $\delta X(C)$ of $X(C)$, from which the innovation would be obtained. Further we wish to have an equation, indeed a variational equation for $\delta X(C)$, which will characterize the probabilistic structure of the random field $X(C)$. Then, we shall come to the problem to solve a stochastic variational equation, and uniqueness of the solution is to be discussed.

To fix the idea and to avoid non-essential complex assumptions, we restrict our attention, in this section, to the case $d = 2$ where the parameter C is in $\mathbf{C}$ containing smooth convex contours (or loops) in the plane. With this choice of $\mathbf{C}$ we propose a stochastic variation equation for $X(C)$.

Recall again that our approach to random fields is based on its innovation. This concept for random fields is to be understood in the following statement. The system $\{Y_s, s \in C\}$ is independent of every $X(C')$ with $C' < C$ and it tells us that the new information which the random field gains between C and $C + \delta C$ should be the same as that gained by Y_s's when s runs through the same region between C and $C + \delta C$.

It is claimed that the stochastic variational equation (8.1.1) if it exists, can determine the probabilistic structure of the given random field $X(C)$ completely, although the formula itself has only a formal significance.

In what follows we assume that

$$X(C) \textit{ is causal} \text{ in terms of white noise.} \tag{8.6.1}$$

Namely, $X(C)$ is a function only of the $x(u)$, $u \in (C)$, (C) being the domain enclosed by C and $x \in E^*$. We are now ready to discuss a random field $X(C)$ satisfying the condition:

$$X(C) = X(C, x) \text{ is in } (S)^* \text{ and homogeneous in } x. \tag{8.6.2}$$

Here homogeneity in x implies that the S-transform $U(C, \xi)$ is a homogeneous polynomial in ξ of degree n in the sense of P. Lévy. In addition, we assume that

$$X(C, x) \text{ is a regular functional,} \tag{8.6.3}$$

that is, the kernel is not a generalized but an ordinary function.

This assumption means that the kernel function which is given by the following proposition is an ordinary $L^2(R^{2n})$-function.

Proposition 8.6 *Under these assumptions* (8.6.1), (8.6.2) *with degree* n *and* (8.6.3), *there is a positive integer* n *such that* $X(C)$ *can be expressed in the form*

$$X(C) = \int_{(C)^n} F(C; u_1, \ldots, u_n) : x(u_1)x(u_2)\cdots x(u_n) : du^n, \tag{8.6.4}$$

where $u_i \in (C), i = 1, 2, \ldots, n$, *and* $F(C, u_1, u_2, \ldots, u_n)$ *is a* $L^2(R^{2n})$-*function symmetric in* u_j*'s and where* : : *is the Wick product.*

As before, we use conventional notation to simplify the expressions $u = (u_1, u_2, \ldots, u_n)$ and $: x(u_1)x(u_2)\cdots x(u_n) :=: x^{n\otimes}(u) :$. Then, the above integral (8.6.4) may be written as

$$\int_{(C)^n} F(C; u) : x^{n\otimes}(u) : du^n. \tag{8.6.5}$$

Note that this integral is different from $X(C)$ given in (6.1.1).

Proof of Proposition 8.6 comes almost straightforward from the above three assumptions.

As for the S-transform $U(C, \xi)$ of the $X(C)$, we refer to the formula in P. Lévy ([47] Iér Part, Chapter 5) by comparing our assumptions.

The formula (8.6.5) can be denoted by

$$\int_{(C)^n} \partial_{u_1}^* \partial_{u_2}^* \cdots \partial_{u_n}^* F(C; u) du^n \tag{8.6.6}$$

by using the creation operators.

Our final assumption is that the kernel

$$F(C; u) \quad \text{and} \quad F'_n(C, u; s) = \frac{\delta F(C, u)}{\delta n}(s) \tag{8.6.7}$$

are continuous in u and in (u, s), respectively.

Definition 8.4 The representation *(8.6.5)* is called a canonical representation in the weak sense if

$$\hat{E}\left[X(C) | X(C'), C' < C_1\right] = \int_{(C_1)^n} F(C, u) : x(u)^{n\otimes} : du^n, \tag{8.6.8}$$

for every $C_1 < C$.

The notation $\hat{E}$ means the weak conditional expectation in the sense of Doob. It is defined by the orthogonal projection of $X(C)$ down to the closed *linear* manifold spanned by the $X(C')$, $C' < C$.

Proposition 8.7 *The representation* (8.6.5) *is a canonical representation in the weak sense if and only if*

$$\int_{(C)^n} F(C; u) f(u) du^n = 0 \tag{8.6.9}$$

for all C with $C < C_1$ implies $f = 0$ on $(C_1)^n$.

Proof. The assumption means that the closed linear manifold generated by the $X(C)$'s with $C < C_1$ is the same as the one generated by all the homogeneous chaos of degree n of the form

$$X(C_1, f) = \int_{(C_1)^n} f(u) : x(u)^{n\otimes} : du^n \tag{8.6.10}$$

with $f \in L^2_{loc}(R^{2n})$. Hence the right hand side of (8.6.10) is just the orthogonal projection of $X(C)$ down to the one generated by the $X(C, f), f \in L^2_{loc}(R^{2n})$. Hence the assertion of the proposition holds. (Cf. [19].)

Let us take the variation $\delta X(C)$ of the field defined by (8.6.5). Then, it is expressed in the form

$$\begin{aligned}\delta X(C) = n \int_C \int_{(C)^{n-1}} & F(C, v_1; s) : x^{(n-1)\otimes}(v_1)x(s) : dv_1^{n-1}\delta n(s)ds \\ + \int_C \int_{(C)^n} & F_n'(C, u; s) : x^{n\otimes}(u) : \delta n(s)du^n ds,\end{aligned} \tag{8.6.11}$$

where $v_1 = (u_2, u_3, \ldots, u_n)$ and F_n' denotes the functional derivative of $F(C; u)$ in the variable C evaluated at C.

Take the weak conditional expectation.

$$\hat{E}(\delta X(C)/X(C'), C' < C) = \int_C \int_{(C)^n} F_n'(C, u)(s) : x^{n\otimes}(u) : \delta n(s)du^n ds. \tag{8.6.12}$$

Then we have

$$\begin{aligned}&\delta X(C) - \hat{E}(\delta X/X(C'), C' < C) \\ &= n \int_C \int_{(C)^{n-1}} F(C, v_1; s) : x^{(n-1)\otimes}(v_1)x(s) : dv_1^{n-1}\delta n(s)ds.\end{aligned}$$

Let δn vary in the class of C^∞-functions so that δC is taken outward and that the integrand over C is determined as a function of s and the right hand side will give

$$x(s) \int_{(C)^{n-1}} F(C, v_1; s) : x^{(n-1)\otimes}(v_1)dv_1^{n-1}. \tag{8.6.13}$$

Let us denote it by

$$x(s)\varphi(s)$$

and use the same technique as in one dimensional parameter space. Thus we know the value

$$\varphi(s)^2.$$

We may ignore its sign to determine $\varphi(s)$. Divide (8.6.13) by $\varphi(s)$ to obtain the generalized innovation $x(s)$. Since the representation is canonical, it can be regarded as the same as the original $x(s)$. It means that it is the real innovation (not in a generalized sense). Thus we can prove the following theorem.

Theorem 8.3 *The innovation for the random field $X(C)$ given by* (8.6.4) *is obtained as*

$$x(s) = \frac{1}{\varphi(s)} \left\{ \frac{\delta X(C) - \hat{E}[\delta X(C)|X(C'), C' < C]}{\delta n}(s) \right\}.$$

Note that if the representation is canonical, then (8.6.13) gives the original white noise $x(u)$ in (8.6.4). However for the non-canonical case, we can see that

$$\begin{aligned} &\delta X(C) - \hat{E}[\delta X/X(C'), C' < C] \\ &\neq n \int_C \int_{(C)^{n-1}} F(C, v_1; s) : x^{(n-1)\otimes}(v_1) x(s) : dv_1^{n-1} \delta n(s) ds. \end{aligned} \tag{8.6.14}$$

Thus, in this case, we are given a generalized innovation which may be different from the original x.

Remark 8.8 *Observe that the situation is somewhat different from the one dimensional parameter case, i.e. Gaussian case. That is*

$$\delta X(C) - \hat{E}(\delta X(C)/X(C'), C' < C) \tag{8.6.15}$$

is orthogonal to $X(C')$ where C' is inside of C, however it may not be independent of the $X(C')$.

Recall that ∂_u is the differential operator defined as $\frac{\partial}{\partial \dot{B}(u)}$, where $\dot{B}(u)$ is now replaced by $x(u)$.

Proposition 8.8 *The kernel function F in* (8.6.4) *is obtained by*

$$F(C, u) = \frac{1}{n!} \partial_{u_1} \partial_{u_2} \cdots \partial_{u_n} X(C), \tag{8.6.16}$$

where $u_1, u_2, \ldots, u_n$ are different.

A mathematical representation $\{X(C)\}$ of a random complex system is important from the viewpoint of our theory, and actually we can manage the representation theory within the frame work of the white noise analysis.

8.7 Examples

Returning to the idea of the innovation approach, our understanding was that as C varies $X(C)$ creates new information expressed in terms of the innovation. Suppose that the innovation is taken to be a white noise. This is often the case. We can therefore assume that $X(C)$ is a function of white noise as we have dealt with so far. The variational calculus for such random fields can also be well established within the frame work of the white noise analysis, where the space $(S)^*$ of generalized white noise functionals can be provided as the basic background.

Thus, the $X(C)$ is a function defined on the space (E^*, μ), where E^* is a space of generalized functions and μ is the Gaussian measure. The $X(C)$ is therefore viewed as a (generalized) functional of x in E^* parameterized by a manifold C in a certain class $\mathbf{C}$ that is chosen suitably.

The so-called S-transform is a powerful tool for the analysis of a field $X(C) = X(C, x)$ as we have often seen so far:

$$(SX(C))(\xi) = U(C, \xi), \quad \xi \in E. \tag{8.7.1}$$

In fact, we can appeal to the classical theory of functional analysis in order to analyse the $U(C, \xi)$. We are interested in the great heritage by those mathematicians like Poincarè, Hadamard, Volterra, Tonelli and Lévy on mathematical theory of functionals, which is quite useful having rephrased in our setup.

For a non-random functional $F(C)$ of C, its variation, if exists, is given by Volterra form (named by P. Lévy) under a reasonable assumption:

$$\delta F(C) = \int_C \frac{\partial F}{\partial n}(s)\delta n(s) ds. \tag{8.7.2}$$

This formula can be applied to the variation of random fields $X(C)$.

If $U(C, \xi), \xi$ being fixed, is the S-transform of a random field $X(C)$, then we can apply the Volterra form to the variation $\delta U(C, \xi)$. We then have a variational formula for $X(C)$ itself. In many cases, we can form the innovation of $X(C)$ from the formula of its variation.

There are good examples in the case of a stochastic process. They are interesting in themselves, and at the same time they are extended to the case of random fields. We therefore begin with innovations for stochastic processes.

Example 8.2 Bilinear case. (See [18], 1980.)
There is a bilinear stochastic differential equation of the form

$$\delta X(t) = -(aX(t) + a')dt + (bX(t) + b')dB(t), \quad a > 0, \ t \geq t_0$$

in the classical expansion, with the initial condition $X(t_0) = C$, where $dB(t)$ is viewed as a Gaussian random measure. If we assume $E[X(t)] = 0$, then $a' = 0$. If the white noise terminology is used, then

$$\begin{aligned} \dot{X}(t) &= -aX(t) + (bX(t) + b')\dot{B}(t), \\ X(t_0) &= C. \end{aligned}$$

Here we note that $\dot{B}(t)$ is independent of $\mathcal{B}_t(X)$. Hence if we set $U(\xi, t) = S(X(t))(\xi)$, then we have

$$\frac{d}{dt}U(\xi, t) = -aU(\xi, t) + (bU(\xi, t) + b')\xi(t), \quad t > t_0,$$

with the initial condition $U(\xi', t_0) = C$.

This is a linear ordinary differential equation in the variable t, and the solution is easily obtained as follows.

$$U(\xi, t) = Ce^{-\int_{t_0}^t (a - b\xi(u))du} + \int_{t_0}^t b'\xi(s)e^{-\int_{s+}^t (a-b\xi(u))du} ds.$$

We can prove that, for every t, $U(\xi, t)$ is in the reproducing kernel Hilbert space with kernel $\exp\left[-\frac{1}{2}\|\xi - \eta\|^2\right]$. Such a Hilbert space will be discussed in Appendix 3. With the help of this space we see that $X(t)$ is living in (L^2) and that it is continuous in t.

If we have the solution $X(t)$ to be *stationary*, then we let t_0 tend to $-\infty$, and let the first term of the above equation disappear so as no remote past exists.

The H_n-component $X_n(t)$ of the stationary process $X(t)$ has the kernel function (symmetric and in $L^2(R^n)$) given by $F_n(t - u_1, t - u_2, \ldots, t - u_n)$, where

$$F_n(u_1, \ldots, u_n) = \frac{1}{n!}b^{n-1}b' \exp\left[-a \min_{1 \leq j \leq n} u_j \chi_{(0,\infty]^n}(u_1, \ldots, u_n)\right], \quad n \geq 1.$$

It is straight forward to prove that $X(t)$ has the covariance function $\gamma(h)$ of the form

$$\gamma(h) = b^{-2} \exp\left[\frac{bb'}{2a}\right] e^{-a|h|}.$$

Proposition 8.9

(1) Best linear predictor for $X(t)$ in Example 8.2 is in agreement with the best non-linear predictor.
(2) The $\dot{B}$ in the given equation is eventually the innovation of $X(t)$.

Proof of this proposition comes from the covariance function and an actual computation.

Example 8.3 (Also see [18] 1980.)
Again we use the classical notation

$$dX(t) = f(t)X(t)dB(t), \quad t \geq 0, \; X(0) = 1,$$

where $f(t)$ is continuous and it never vanishes. We therefore have

$$\dot{X}(t) = \partial_t^* f(t)X(t).$$

Then with the help of the S-transform again, the unique solution is expressed in the form

$$X(t) = \exp\left[\int_0^t f(u)dB(u) - \frac{1}{2}\|f\|_t^2 du\right],$$

where $\|f\|_t^2 = \int_0^t f(u)^2 du$.
The innovation is $\dot{B}(t)$, with $dB(t)$ appeared in the given equation. It can be obtained in such a way that

$$[\log X(t) - E \log X(t)] = Y(t),$$

which is an additive Gaussian process. Then, $\dot{B}(t)$ is obtained as the innovation of $Y(t)$.

Example 8.4 Define a stochastic process

$$X(t) = B(t)^2 - t.$$

Then we have

$$dX(t) = 2B(t)dB(t), \quad t \geq 0, \; X(0) = 0.$$

This is a sort of stochastic differential equation. It is easy to see that $dB(t)$ in the above expression is not the innovation. It has a unique solution but unfortunately not a Markov process. (See the Itô formula.)

It is well known that $X(t)$ is a martingale, and the innovation in the weak sense is formed by $\frac{B(t)}{|B(t)|}\dot{B}(t)$.

Example 8.5 Set

$$X(t) = B(t)(P(t) + 1).$$

Then, it can be shown, again by observing sample functions, that

$$\mathbf{B}_t(X) = \mathbf{B}_t(B) \vee \mathbf{B}_t(P)$$

holds for every t. Further, it is easy to prove that

Proposition 8.10 *$X(t)$ is a martingale and the innovation is*

$$\dot{B}(t) + \dot{P}(t).$$

Example 8.6 Kailaith's example.

Let $X(t), t \geq 0$, be given by

$$X(t) = S(t) + w(t),$$

where $S(t)$ is a signal and $w(t)$ is a noise. We assume that the signal process $S(t)$ is a Gaussian process with $ES(t) \equiv 0$ and $w(t)$ is a white noise which is independent of the signal process. Then $X(t)$ is again Gaussian process with $EX(t) = 0$. Further the covariance $\Gamma(s,t) = E[X(s)X(t)]$ is continuous in (s,t), hence so is for $E[S(s)S(t)]$ $s \neq t$.

Set

$$\hat{S}(t) = E[S(t)|X(s), s \leq t].$$

(Note that the condition does not contain the information of $w(t)$.) Then, we can prove that the difference

$$\tilde{S}(t) = S(t) - \hat{S}(t)$$

is the innovation.

Actually, if we are allowed to express

$$\tilde{S}(t) = \int_0^t h(t,s)X(s)ds,$$

with a kernel $h(t,s), s < t$, then h is a solution of the following integral equation:

$$h(t,s) + \int_0^t h(t,u)\Gamma(u,s)du = \Gamma(t,s), \quad s < t.$$

Generalizations of the above result are also known.

Remark 8.9 *A. N. Shiryaev established the similar results, independently, atmost the same time.*

Example 8.7 Let $X(t)$ be given by

$$X(t) = \varphi(B(t))\psi(P(t)), \quad t \geq 0,$$

where φ is a homomorphism between R and the range of φ, and where ψ is an increasing function on $[0,\infty)$ with $\psi(0) > 0$.

Then the following relationship can be proved:

$$\mathbf{B}_t(X) = \mathbf{B}_t(B) \vee \mathbf{B}_t(P), \quad \text{for every } t.$$

With this property, it is proved that $X(t)$ has an innovation $a\dot{B}(t) + b\dot{P}(t)$ with arbitrary non-zero constants a and b. This is a generalization of the result in Section 8.4.

Let $B(t)$ be a Brownian motion and $P(t)$ be a Poisson process. $X(t) - \int_0^t (2B(u)P(u) - B(u))du$ is proved to be a martingale. We can see that by using the stochastic derivative of $X(t)$, the innovation of $X(t)$ is given.

Example 8.8 R^2 parameter Lévy's Brownian motion.

It is known that $\{\frac{1}{\sqrt{2\pi}}, \sqrt{\pi}\cos k\theta, \sqrt{\pi}\sin k\theta, k \geq 1\}$ is a complete orthonormal system in S^2. Denote it by $\{\varphi_n(\theta), n \geq 0\}$. The Lévy Brownian motion with $X(a), a \in R^2$ can be written as $X(r,\theta)$, $a = (r,\theta)$, by the polar coordinates.

Set

$$X_n(t) = \int_0^{2\pi} X(t,\theta)\varphi_n(\theta)d\theta, \quad n \geq 0. \tag{8.7.3}$$

We can see that $\{X_n(t), n \geq 0\}$ is an independent system since the covariance

$$E[X_m(t)X_n(s)] = 0, \quad m \neq n,$$

for any s and t.

Let the representation of $X_n(t)$ for 2-dimensional parameter case be

$$X_n(t) = \int_0^t F_n(t,u)\dot{B}(u)du, \quad n \geq 0.$$

By using McKean's expansion (see [56]) the canonical kernel $F_n(t,u)$'s of the representation is obtained as

$$\begin{aligned} F_0(t,u) &= \frac{1}{\pi}\cos^{-1}\frac{u}{t} \\ F_n(t,u) &= \frac{1}{2\pi}\left(\frac{u}{t}\right)^{n-1}\left(1-\frac{u^2}{t^2}\right)^{1/2}, \quad n > 0. \end{aligned}$$

Set

$$F_n(t,f) = \int_0^t F_n(t,u)f(u)du, \quad f \in C\{[0,\infty)\}.$$

Note that

$$t^l F_n(t,f) = \frac{1}{4\sqrt{\pi}}\int_0^{t^2} v^{(l/2)-1}\sqrt{t^2-v}\,f(\sqrt{v})dv,$$

and then we can prove the formula

$$\Gamma\left(\frac{3}{2}\right)(D_{t^2})^{3/2}\,t^l F_n(t,f) = \frac{1}{2}t^{l-2}f(t).$$

Thus we have

$$L_t^{(n)}F_n(t,f) = f(t),$$

and so

$$L_t^{(n)}X_n(t) = \dot{B}_n(t) \tag{8.7.4}$$

where

$$L_t^{(n)} = \sqrt{\pi}t^{-l+2}D_{t^2}^{3/2}\,t^l, \quad D_t = \frac{d}{dt}. \tag{8.7.5}$$

That is the way how the white noise $\dot{B}_n(t)$ can be formed from $X_n(t)$, with $n = 2l-1,\ l > 0$ as

$$L_t^{(l)}X_n(t) = \dot{B}_n(t). \tag{8.7.6}$$

It can be easily seen that $\{\dot{B}_n(t)\}$ is an independent system of white noises since $\{X_n(t)\}$ is an independent system.

The situation is similar for the case $n = 2l$. It should be noted that for even dimensional parameter case, the innovation is obtained by applying the operator $L_t^{(l)}$, to Lévy's Brownian motion, which is not local, but causal.

Example 8.9 We now consider R^3 parameter case. The canonical kernel function of X_n is known as

$$F_0(t,u) = 1 - \frac{u}{t},$$
$$F_n(t,u) = \frac{1}{2}\left(\frac{u}{t}\right)^{n-1}\left(1 - \frac{u^2}{t^2}\right), \quad n > 0.$$

Note that, in the expression of the Brownian motion, there appear $2n+1$ independent processes with the same canonical kernel F_n.

It is easy to show that there exists a local operator

$$L_t^{(l)} = \frac{1}{t^{l-1}} \frac{d}{dt} \frac{1}{2t} \frac{d}{dt} t^{l+1}, \tag{8.7.7}$$

such that

$$L_t^{(l)} X_n(t) = \dot{B}_n(t).$$

Remark 8.10 *For odd dimensional parameter case, say R^{2p+1}-parameter case, $M(t)$ process is $(p+1)$-ple Markov process as is well known.*

Example 8.10 Gaussian martingale. (Cf. Section 5.3.)

Let $X(C), C \in \mathbf{C}$ with $d = 2$ be given by a stochastic integral of a locally square integrable g:

$$X(C) = \int_{(C)} g(u)x(u)du^2.$$

Then, with the help of the S-transform it is easy to see

$$\delta X(C) = \int_C g(s)x(s)\delta n(s)ds, \tag{8.7.8}$$

where ds is the line element over C.

Conversely, if we are given a variational equation of the above form, it is easy to see the solution which is in agreement with $X(C)$ up to constant.

Example 8.11 The variational equation of Langevin type.

The Langevin equation for a Gaussian random field is a field defined as follows. Let $\mathbf{C}_0$ be a class of plane circles and G be the conformal group $C(2)$

$$\delta X(C) = -X(C)\int_C \varphi(s)\delta n(s)ds + X_0 \int_C \nu(s)x(s)\delta n(s)ds, \quad C \in \mathbf{C}_0,$$

where ϕ and ν are given continuous functions.

Apply the S-transform and denote the S-transform of $X(C)$ by $U(\eta) = U(\eta, \xi), \xi$: fixed where η denotes an analytic representation of C. Then

$$\delta U(\eta)(\xi) = \int \varphi(s)\delta\eta(s)ds + U_0 \int v(s)\xi(s)\delta\eta(s)ds.$$

Hence, we obtain the solution

$$X(C) = X_0 \int_{(C_0)} \exp[-k\rho(C,u)]\partial_u^* \nu(u)du,$$

where ρ denotes the Euclidian distance. The solution of such a Langevin type stochastic variational equation is a simple Markov Gaussian random field.

As a generalization of Example 8.11, there is a stochastic variational equation for $X(C)$ is given by

$$\delta X(C) = -aX(C)\int_C \delta n(s)ds + b : X(C)^n : \int_C x(s)\delta n(s)ds,$$

where x is a sample function of white noise with d-dimensional parameter white noise.

Example 8.12 A random field expressed as a homogeneous chaos.
This has been discussed in Section 6.1.

Example 8.13 Let C be in $\mathbf{C}$ as usual, and let $G(u, v, C), (u, v) \in R^2$, be the Green function associated with the two dimensional Laplacian operator $\Delta = \Delta_u = \frac{\partial^2}{\partial u_1^2} + \frac{\partial^2}{\partial u_2^2}$. Define a random field $X(C, u)$ with parameter (C, u):

$$X(C,u) = \int_{(C)} G(u, v, C)x(v)dv.$$

Then

$$\Delta_u X(C, u) = x(v).$$

Although $x(v)$ is a generalized function, but a rigorous interpretation is given by the S-transform. Thus, the *reduction* is established. While, the variation in the variable C is given by

$$\delta X(C,u) = \int_{(C)} \int_C \frac{\partial G(u,m;C)}{\partial n(m)} \frac{\partial G(m,v;C)}{\partial n(m)} x(u) \delta n(m) dm\, du,$$

where the Hadamard equation is employed (see [15]).

Example 8.14 General Gaussian processes.

First recall Theorem 8.1 which shows a method how to construct the innovation of a Gaussian process satisfying minor assumptions.

Let $X(t)$ be a separable Gaussian process having no remote past. Then the general theory for the canonical representation (see [1], Part II, [8]) asserts that there exists atmost countably infinite number of cyclic subspaces with cyclic vectors X_m, defined by the family of projection $\{E(t), t \in R\}$.

Now assume there is no point spectrum of $E(t)$. Then, each $E(t)X_n = X_n(t)$ is a Gaussian additive process with unit multiplicity. Thus, the collection $\{dX_n(t)\}$ forms the innovation.

Example 8.15 General Gaussian processes.

We finally come to an investigation of the system $\{X(C); C \in \mathbf{C}\}$ by viewing it as a random complex system. In a suitable manner a linear order is introduced for a subclass of $\mathbf{C}$. According to the deformation of the C within this linearly ordered subset, one can consider the multiplicity of $X(C)$ as C proceeds. This concept can express the degree of complexity of the system. Many interesting examples have countably many multiplicity of Lebesgue type. (Cf. Kakutani's pioneering work : see [18], Chapter 1.)

A randomized version can be introduced and its innovation is obtained.

Chapter 9

Reversibility

The reversibility is an important characteristic property of a random field, since we regard it as a mathematical model of an evolutional phenomenon which is reversible. This section has close connection with classical mechanics as well as quantum dynamics. First, we shall consider the case of a stochastic process, then some results on a reversible random field will be briefly mentioned.

9.1 Reversibility of stochastic processes

There are two important directions, among others, in the study of evolutional random complex phenomena, as for the dependence on the parameter. Namely, we claim

(1) Markov property and multiple Markov property (this can be well defined only for Gaussian case), and
(2) Reversibility.

Both are defined in connection with the development of the time or space-time, and the two are related with each other, although they are discussed by different tools from analysis. We shall first consider the case of a stochastic process $X(t)$. In particular, we discuss Gaussian and Poisson noises. Later we shall come to a random field $X(C)$.

(i) Gaussian case

Some observation will be made in the simplest case where the given stochastic process $\{X(t), 0 \le t \le 1\}$ is a Gaussian Markov process.

Suppose $X(t)$ is a Brownian bridge such that $X(0) = 0, X(1) = 0$, and $X(t), 0 < t < 1$ fluctuates like a Brownian motion. It may be obtained from

Brownian motion by taking the bridged effect into account:

$$X(t) = B(t) - tB(1).$$

Its covariance function is

$$\Gamma(t,s) = (t \wedge s)((1-t) \wedge (1-s)),$$

which tells us that $\{X(t)\}$ and $\{X(1-t)\}$ have the same probability distribution. In this sense we can see the reversibility.

More profoundly, we can illustrate the reversibility in terms of their sample paths. Namely, based on a white noise we can give the canonical representation, which is now viewed as a sample function-wise integral:

$$X(t) = \int_0^t \frac{1-t}{1-u} \dot{B}(u) du.$$

Since $\dot{B}(t)$ and $\dot{B}(1-t)$ have the same probability distribution, the canonical representation of reversed process is

$$X_1(t) = \int_t^1 \frac{t}{v} \dot{B}(v) dv.$$

Because of the reversibility in this sense just explained, a Brownian bridge appears in the representations of many reversible random complex phenomena. For example, the amount of fluctuation of a classical trajectory to obtain a quantum mechanical path is expressed by the Brownian bridge up to constant. This fact is essential when we transfer classical dynamics to Quantum mechanics. Thus, the idea is applied to the Feynman path integral in [83], where fluctuating (possible) trajectories contribute to have the propagator of the quantum dynamics in question.

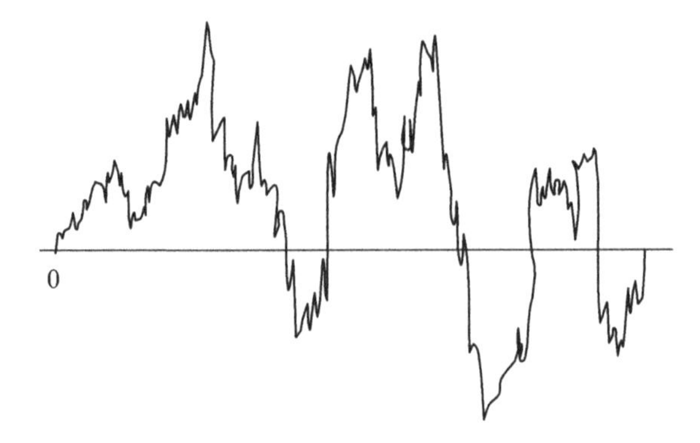

Fig. 9. Brownian bridge.

As a generalization of the above transformation, we can consider a *conformal* transformation $t \to \frac{t-1}{at-1}$ with a parameter a, $a < 1$. This fact is also suggested by the following observation.

Let $X(t)$ above be normalized; that is, let us set

$$Y(t) = \frac{X(t)}{\sigma(t)},$$

where

$$\sigma^2(t) = E[X(t)^2].$$

The covariance function of $Y(t)$ is given by

$$\gamma(t,s) = \sqrt{\frac{s/t}{(1-s)/(1-t)}} = \sqrt{\frac{s(1-t)}{t(1-s)}},$$

which is the anharmonic ratio $(0,1;s,t)$. This immediately implies the *projective invariance* of $Y(t)$ (or of Brownian motion). This fact is explained as follows.

Let $p(t), t \in [0,1]$ be a projective transformation of $[0,\ 1]$ onto itself. Then $\{Y(t), t \in (0,1)\}$ and $\{Y(p(t)); t \in (0,1)\}$ have the same probability distribution.

It is easy to see that the covariance function of $\{Y(p(t)), t \in (0,1)\}$ is equal to $\gamma(t,s)$.

There is an important note. What we have claimed is that the projective invariance is realized by changing the variables of sample functions, not simply the invariance of the probability distribution.

It is worth to be mentioned that the following assertion holds.

Theorem 9.1 *The Brownian bridge $X(t)$ over the interval $[0,1]$ is characterized (up to constant) by the conditions*

(i) *$X(t)$ is a Gaussian Markov process that has the canonical representation,*
(ii) *$X(0) = X(1) = 0$ (bridged),*
(iii) *the normalized process $Y(t)$ enjoys the projective invariance,*
(iv) *the local continuity of $Y(t)$ as $t \to 0$ in terms of the covariance function is the same as that of the normalized Brownian motion $B(t)/\sqrt{t}$.*

A generalization to multi-dimensional parameter case will be discussed in the next section, where one can see an important connection between rotation group and white noise analysis. Thus, the latter has an aspect of harmonic analysis arising from the infinite dimensional rotation group. A similar observation is also given in the next section.

(ii) Poisson case

Consider the case $d = 1$, and remind our method to restrict our attention to the event $A(n)$, where $P(1,\omega) = n$. Define $Q(t,\omega) = n - P(1-t,\omega)$. Then, $P(t), 0 \le t \le 1$, has the same probability distribution as $Q(t), 0 \le t$, on the event $A(n)$. Thus, we claim a reversibility of Poisson noise only by restricting ω to be in $A(n)$. This means reversibility.

In terms of noise, we take time derivatives of two processes; $\dot{P}(t)$ and $\dot{Q}(t)$ which is $\dot{P}(1-t)$. The conditional (that is on the event $A(n)$) characteristic functionals are

$$\left(\int_I e^{i\xi(t)}dt\right)^n$$

and

$$\left(\int_I e^{i\xi(1-t)}dt\right)^n,$$

respectively, where $I = [0,1]$. The two are equal, which means the reversibility from a particular viewpoint.

9.2 Reversibility of a random field

(1) Gaussian case

We have been tacitly thinking of a way of generalizing the reversibility to the case of random field $X(C)$, where the parameter C, which is an ovaloid in R^2, runs through C_0 to C_1 with $(C_0) \subset (C_1)$. The way of deformation of C would be a generalization of the transformation $t \to 1-t$.

For this purpose, we recall the special case of linear parameter. The reflection $t \to 1-t$ is a particular transform of a conformal transformation of [0, 1] onto itself in a reversible order. This having been generalized, we are given a slight generalization to consider a conformal transformation from $t \to \frac{t-1}{at-1}$ with parameter a $(a < 1)$.

Thus, we are ready to introduce a class G of conformal transformations that deforms C_1 to C_0 monotonically in a manner $C \to gC$ with $gC_0 = C_1, gC_1 = C_0$ and $g \in G$.

On the other hand, the conformal transformations define a subgroup $C(2)^*$ of $O^*(E^*)$. This means that $g^* \in C(2)^*$ keeps the white noise manner invariant $(g^*\mu = \mu)$.

Hence it is quite reasonable to introduce the way of deformation gC of C in such a way that gC runs through C_1 to C_0 as C deforms from C_0 to C_1.

We now give the analogue of the Brownian bridge for the Gaussian random fields, discussed above. Let C_r be a circle with radius r, and take C_0 and C_1 such that $C_0 < C_r < C_1$. By writing $f(C_r)$ by $f(r)$, the representation of $X(C_r)$ can be written as

$$X(C_r) = f(r) \int_{(C_r)-(C_0)} g(u)x(u)du^2,$$

where

$$f(r) = \sqrt{\frac{(r_1^2 - r_0^2)(r_1^2 - r^2)}{r^2 - r_0^2}}$$

and

$$g(u) = (r_1^2 - |u|^2)^{-1},$$

where r runs through the interval $[r_0, r_1]$.

We can see that the above $X(C)$ is a random field like a Brownian bridge such that $X(C_0) = X(C_1) = 0$. Thus we may call it a bridged random field. The covariance function is obtained as

$$\Gamma(s,t) = \pi \sqrt{\frac{(s^2 - r_0^2)(r_1^2 - t^2)}{(r_1^2 - s^2)(t^2 - r_0^2)}}.$$

Note that $\Gamma(s,t)$ is the square root of the anharmonic ratio $(s^2, t^2; r_0^2, r_1^2)$ up to a constant π. Hence, one can think of the invariance of the field in question under the projection transformations in the variable u^2.

By applying a conformal transformation to white noise as

$$x(u) \to x\left(\frac{r_0 r_1}{|u|^2} u\right) \frac{r_0 r_1}{|u|^2},$$

we obtain the random field

$$X(C_r) = \sqrt{\frac{(r_1^2 - r_0^2)(r^2 - r_0^2)}{r_1^2 - r^2}} \int_{(C_{r_1})-(C_r)} (|v|^2 - r_0^2)^{-1} x(v) dv^2,$$

which has the same covariance function with that of the original random field $X(C)$ since the above transformation keeps the white noise measure invariant. Thus we see the reversibility of the random field.

The characteristics of this $X(C)$ are the same as in the case of Brownian bridge.

It is straightforward to have generalization of reversible fields formed by higher dimensional parameter white noise. In fact, in the R^d-parameter case, functions $f(r)$ and $g(u)$ have the same expression simply by replacing monomials like r^2, u^2 with r^d, u^d, respectively.

As regards the Gaussian field $X(C_r)$, it is easy to have another generalization by using a diffeomorphism g acting on R^2 in such a way that $gX(C_r) = X(gC_r)$.

By taking general $C \in \boldsymbol{C}$, instead of $X(C_r)$, we are given a Markov random field and we have a stochastic variation that uniquely determines the $X(C)$.

(2) Poisson case

By replacing Gaussian white noise with Poisson noise $V(u)$, a sample function of which is denoted by $y(u), u \in R^2$, we are given a reversibility random field $Y(C)$. Set

$$Y(C_r) = f(r) \int_{(C_r)-(C_0)} g(u)y(u)du^2, \tag{9.2.1}$$

where $f(r)$ and $g(u)$ are the same functions as those in the Gaussian case, respectively.

Now apply a dilation to the parameter u of $V(u)$:

$$\tau_t : u \;\rightarrow\; ue^{at}, \quad a \geq 0.$$

Let $(U_t y)(u) = y(\tau_t u)e^{2at}$, then U_t is measurable. Thus, $Y(C_r)$ is transformed to be $Y_t(C_r)$:

$$\begin{aligned} Y(C_r) &= f(r) \int_{(C_r)-(C_0)} g(u)y(ue^{at})e^{2at}du^2 \\ &= f_t(r) \int_{(C_{e^{at}r})-(C_{e^{at}r_0})} \frac{1}{(e^{at}r_1)^2 - |v|^2} y(v)dv^2 \end{aligned}$$

where $f_t(r)$ is obtained by replacing $\{r_0, r, r_1\}$ with $\{r_0e^{at}, re^{at}, r, e^{at}\}$, respectively.

According as t varies positive or negative direction, the random field $Y(C_r)$ moves back and forth keeping the same structure as a Poisson noise functional. So, one may observe reversible phenomenon. Note that U_t is not μ_P measure preserving. In fact, intensity λ changes to λe^{2at} under U_t.

9.3 Variational equations for quantum fields

There are so many interesting questions concerning reversibility in connection with the quantum probability theory. It is almost impossible to explain each question, so that only some of them will be shown below.

(A) Reversibility of general $X(C)$

Analogous approach to the case of one-dimensional time t needs to introduce an orientation in the class $\mathbf{C}$. Thus "future and past", lie in the "reverse order", and hence the notion of "path" etc. can be introduced.

paths of $X(C) \to$ path integral where the path is in $\mathbf{C}^{[0,1]}$.

Each sample function of $X(C)$ is viewed as a path of the random quantity $X(C)$ of the variable C. Thus a probability measure is naturally introduced in $\mathbf{C}^{[0,1]}$.

(B) Approach to quantum fields

We simply list the possible directions in quantum dynamics that can be approached by our method.

Tomonaga–Schwinger equation (see Section 10.7),

Nambu–Goto Strings (see [97]),

The Ostwalder–Schrader axioms for Euclidean field will lead to a quantum field (see Section 5.6),

Quantum optics (see [4]), path integral formulation to have propagator in line with Feynman's theory,

Quantum geometry of Bosonic string (see A. M. Polyakov, Phys. Letters 103 B (1981), 207–210),

Yang–Mills theory: It is concerned with the singularity on the diagonal of quadratic generalized white noise functionals, although we have briefly discussed.

Remark 9.1 *Remark on a generalization. For a curve $C \in \mathbf{C}$, if there exists one parameter family of conformal transformations that connects C_0 and C_1, the reversibility can be considered in a similar manner. This will be done by a method of differential geometry. In this case $\{X(C_r), r_0 \le r \le r_1\}$ is viewed as a path of $X(C)$. Its variation is expressed as $d_r X(C_r)$.*

Chapter 10

Applications

10.1 Conformal group $C(d)$ as a subgroup of $O(E)$

The conformal group $C(d)$ is going to be introduced as the most important subgroup of the $O(E)$. Since it involves the reflection with respect to the unit sphere, it is therefore necessary to take the basic nuclear space E to be the space D_0 for which the transformation defined by the reflection is its diffeomorphism of E. Actually the space $D_0 = D_0(R^d)$ for the d-dimensional parameter case is defined by

$$D_0(R^d) = \{\xi \in C^\infty(R^d);\ s\xi \in C^\infty\},$$

where $s\xi(u) = \xi\left(\frac{u}{|u|^2}\right)\frac{1}{|u|^d}$. The topology of $D_0(R^d)$ is introduced so as to be isomorphic to $C^\infty(S^d)$, S^d being the d-dimensional unit sphere. This space $D_0(R^d)$ is fitting for our purpose.

First, we consider the one-dimensional parameter case. The isomorphism between $D_0(R^1)$ and $C^\infty(S^1)$ is now given by

$$\xi(u) \to \tilde{\xi}(\theta) = \sqrt{2}\xi\left(\tan\frac{\theta}{2}\right)\cos\frac{\theta}{2}.$$

Let $\tilde{G}$ denote the projective special linear group $PSL(2, R)$. A member $\tilde{g}$ of $\tilde{G}$ is represented by a 2×2 matrix:

$$\tilde{g} = \begin{pmatrix} a & b \\ c & d \end{pmatrix}, \tag{10.1.1}$$

with $ad - bc = 1$ and modulo center.

The $\tilde{g}$ defines an operator g acting on $E = D_0$, as a continuous linear operator, in such a way that

$$\xi \ \to g\xi(u) = \xi\left(\frac{au+b}{cu+d}\right)\frac{1}{|cu+d|} \qquad u \in R.$$

Obviously g is continuous in the topology of E and it preserves the $L^2(R)$-norm. That is, g is a member of $O(E)$.

The collection of such g's forms a group, actually a closed subgroup of $O(E)$, denote it by G.

We refer to whiskers in Section 2.4(4).

Triviality There is an isomorphism

$$G \cong PSL(2, R),$$

between the two topological groups.

It is known that the G is generated by the following subgroups:

(1) Shift S_t: $\xi(u) \to \xi(u-t), \quad t \in R,$

(2) dilation τ_t: $\xi(u) \to \xi(ue^t)e^{t/2}, \quad t \in R,$

(3) special conformal transformations: $\kappa_t = wS_tw, t \in R$, and the reflection w, where

$$w\colon \xi(u) \to \xi(1/u)\frac{1}{|u|^2}.$$

In the higher dimensional case, say d-dimensional case ($d > 1$), an additional subgroup, the rotation group $SO(d)$, is added in the above list. Namely, a rotation $\tilde{g}_\theta$ in the group $SO(d)$ defines a rotation g_θ of E:

(4) rotations $g_\theta : \xi(u) \to \xi(\tilde{g}_\theta u), \tilde{g}_\theta \in SO(d)$.

In addition, d shifts and d special conformal transformations are included in G.

The dilation should be the isotropic dilation in the case of R^d, so that only one dilation is available. If non-isotropic dilation is included, then the Lie group generated by others in our list becomes infinite dimensional, so that we want to avoid such a case.

The subgroup of $O(D_0(R^d))$ thus obtained is often called the d-dimensional *conformal group* and is denoted by $C(d)$. Each one-parameter subgroup listed above is a *whisker*. (See the figure of the $O(E)$ Fig. 3, where each one-parameter subgroup passing through the identity e looks like a whisker.)

Recall the role of the infinite dimensional rotation group to see:

Triviality The collection $C(d)^* = \{g^*;\ g \in C(d)\}$ forms a subgroup of $O(D_0)^*$, and each one-parameter subgroup of $C(D_0)^*$ which involves adjoint members of a whisker is a *flow* on the measure space (D_0^*, μ).

Proposition 10.1 *The subgroup $C(d)$ of $O(D_0(R^d))$ is locally isomorphic to the Lie group $SO(d+1,1)$.*

Proof. First compute the infinitesimal generators of the one-parameter groups listed above (1)–(4). They are

(1) shift: $s_j = -\frac{\partial}{\partial u_j}, 1 \le j \le d$,
(2) isotropic dilation: $\tau = u\frac{d}{du} + \frac{1}{2}$,
(3) special conformal transformations: $\kappa_j = u_j^2 \frac{\partial}{\partial u_j} + u_j$,
(4) rotations: $\gamma_{j.k} = u_j \frac{\partial}{\partial u_k} - u_k \frac{\partial}{\partial u_j}, 1 \le j \ne k \le d$.

Form the Lie algebra generated by these generators to see that the algebra is isomorphic to that of $SO(d+1,1)$ (often denoted by $\tilde{C}(d)$).

Remark 10.1 *The generators (1), (2) and (3) define whiskers. There are $2d+1$ whiskers and they play different roles in probability, respectively.*

The above generators are used in case the variational calculus for $X(C)$ is obtained, where infinitesimal deformation of the parameter C is performed by the conformal transformations. (See Section 10.4.)

10.2 Spectral type of flows

We are given various flows, so that it would be fine if their spectral types are clarified.

First, the shift is observed. The flow $\{T_t\}$ with $S_t^* = T_t$ is called the *flow of Brownian motion* after S. Kakutani. Associated with the flow $\{T_t\}$ is a one-parameter unitary group U_t defined by $(U_t\varphi)(x) = \varphi(T_t x)$ for $\varphi \in (L^2)$. Now use the Fock space:

$$(L^2) = \bigoplus_n H_n,$$

and take the integral representation of the members in H_n by applying the S-transform. To fix the idea, we take unitary group U_t acting on H_2. The S-transform gives us a symmetric $L^2(R^2)$-function $F(u,v)$ and the effect of U_t implies

$$F(u,v) \to F(u+t, v+t).$$

Changing the variables: it suffices to consider the half plane $u \ge v$, so

$$x = u+v, \qquad y = u-v \ (\ge 0),$$

to have

$$F(u,v) = \tilde{F}(x,y), \qquad F(u+t, v+t) = \tilde{F}(x+2t, y).$$

Choose an arbitrary base $\{\eta_n\}$ of $L^2([0,\infty)]$ to have an expansion of F

$$\tilde{F}(x,y) = \sum_n f_n(x)\eta_n(y).$$

Each term has the second factor that is invariant under the shift. This proves that the shift spans cyclic subspaces, which are mutually orthogonal, and each of which has the Lebesgue spectrum. Thus, we have proved that the U_t and hence T_t has countably Lebesgue spectrum.

The same for the subspace $H_n, n \geq 3$.

Summing up, we conclude that

Theorem 10.1 *The flow of Brownian motion has countable Lebesgue spectrum on the space* $(L^2) \ominus \{1\}$.

Spectrum of the dilation. It is countably Lebesgue, too.

For more details concerning the spectrum we refer to [18, Chapter 5].

We then come to the spectrum of flows defined by the shifts in the R^d-parameter case. We know that as a subgroup of $C(d)$ we are given shifts, denote them by $S_t^j, j = 1, 2, \ldots, d$, as many as d. Hence we are given d flows T_t^j.

Proposition 10.2 *Each flow* T_t^j *defined by a shift* S_t^j *has countable Lebesgue spectrum on the space* $(L^2) \ominus \{1\}$.

The idea of the proof is a simple modification of the case $d = 2$. Namely, one of the variables of the kernel function is shifted by t and other variables are kept invariant. We can then play the same game as in Theorem 10.1.

One may now ask what are the relations among those shifts or flows. To answer this question, one may take the infinitesimal generators s_j of S_t^j, respectively. Obviously, we prove

(i) s_j's are commutative, and
(ii) s_j's are linearly independent in the Lie algebra $c(d)$ associated with the conformal group $C(d)$.

In general, the notion of *multiplicity* shows complexity of the dynamical system. In terms of the shift operators, we have shown part of the complexity of white noise by (i) and (ii) together with the property of countable Lebesgue spectrum.

10.3 The transversal relation

There is a good relationship between the shift and dilation. Namely the shift is *transversal* to the dilation, which may be expressed in terms of the commutation relation of the generators. In fact, it is easily proved that

$$S_t^j \tau_s = \tau_s S_{te^s}^j, \quad j = 1, 2, \ldots, d.$$

In terms of the generators τ and s_j we have

$$[\tau, s^j] = -s^j, \quad j = 1, 2, \ldots d.$$

It is interesting to note that the shift defines the flow of Brownian motion, while the dilation defines the flow of Ornstein–Uhlenbeck process.

This relation is used when dynamical property of the flows defined by the shift and dilation is investigated.

The transversal relations can also be seen for other pairs, like

$$[\tau, \kappa_j] = \kappa_j, \quad j = 1, 2, \ldots, d.$$

Such a relation is helpful for the study of ergodic property of flows on white noise space.

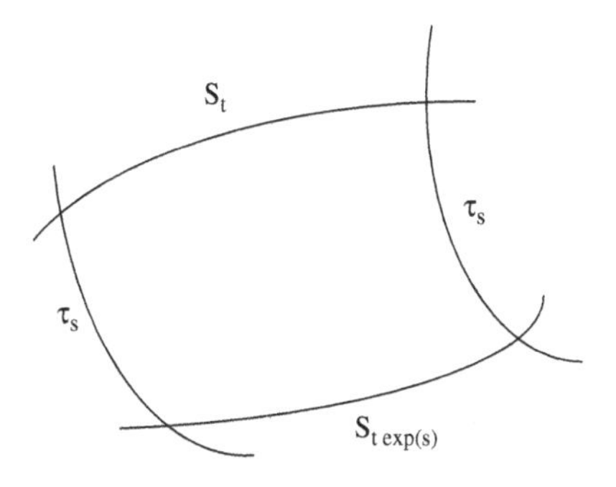

Fig. 10. Transversal relation.

10.4 Conformal invariance of white noise

(1) In the one dimensional parameter case, we take a Brownian bridge $X(t), t \in [0,1]$, introduced in Section 9.1 to show the reversibility. Let it be normalized to have

$$Y(t) = \frac{X(t)}{\sqrt{t(1-t)}}.$$

The covariance function of $Y(t)$ is the square root of the anharmonic ratio $(0,1;s,t)$, which proves that the probability distribution of $Y(t)$ is invariant under any projective transformation that carries $[0,1]$ to another interval. If the mapping carries $[0,1]$ onto itself, we are given one-dimensional subgroup of the conformal group and the action is visualised. Some details have been observed in the last section.

(2) If we come to the higher dimensional parameter white noise, then it is very much complicated to explain the conformal invariance in an explicit formula, although the invariance itself is quite important. We do not go into details, but refer the paper

"T. Hida, K.S. Lee and S.-S. Lee, Conformal invariance of white noise, Nagoya Math. J. 98 (1985), 87–98,"

that discusses the conformal invariance.

Reversibility, discussed in a similar manner to the case $d = 1$, has been mentioned before.

10.5 Action on random fields

Another role of the whiskers is to have a deformation of the manifold C which is taken to be a parameter of a random field.

Given a Gaussian random field $\{X(C); C \in \mathbf{C}\}$, where $\mathbf{C}$ is a collection of manifolds in a Euclidean space. Then, we are interested in the dependence of $X(C)$ when C moves and deforms within the class $\mathbf{C}$.

Two particular cases will be discussed.

[1] The class of manifolds is chosen to be $\mathbf{C}_0$ the collection of all $(d-1)$-dimensional spheres in R^d. In an obvious manner $\mathbf{C}_0$ may be identified with $R^d \times R_+$ as a topological space.

We recall that there is a conformal group, denoted by $\tilde{C}(d)$, that is acting on R^d, and is consisting of the following (i)–(iv):

(i) shifts,
(ii) isotropic dilation,
(iii) $SO(d)$,
(iv) special conformal transformations = conjugates to the shifts with respect to ω, where ω is the reflection: $u \to \frac{u}{|u|^2}$.

Put the transformation (i)–(iv) together. And one is given the conformal group $\tilde{C}(d)$ which is $\frac{1}{2}(d+1)(d+2)$ dimensional. Its Lie algebra is determined by Proposition 10.1.

The following assertion can easily be proved.

Proposition 10.3 *The class $\mathbf{C}_0$ of spheres is invariant under the action of the conformal group $\tilde{C}(d)$, and the action of the group on the space $\mathbf{C}_0$ is continuous and transitive.*

With this property of the conformal group, we can speak of the variation of a random field depending on a sphere. Set

$$X(C) = \int_C F(s)X(s)dv(s), \quad C \in \mathbf{C}, \tag{10.5.1}$$

where $\{X(s); s \in C\}$ is the restriction to C of a continuous Gaussian random field $X(t), t \in R^d$, $F(s)$ is a continuous function and $dv(s)$ is the surface element over the sphere C.

Infinitesimal deformation δC of C is induced by infinitesimal changes of members in $\tilde{C}(d)$ and eventually it gives us the variation of $X(C)$. Hence, we have to consider the action of the Lie algebra of $\tilde{C}(d)$. Let $C(d)$ be regarded as a continuous representation of $\tilde{C}(d)$ on E; namely for $\tilde{g} \in \tilde{C}(d)$

$$g\xi(u) = \xi(\tilde{g}u)|J|^{1/2}, \quad u \in R^d, \ J: \text{Jacobian}.$$

We can take a base $\{\alpha_j; 1 \le j \le \frac{1}{2}(d+1)(d+2)\}$ of the Lie algebra of the group $C(d)$. Members of the base may come from one-parameter subgroups (whiskers) $g_t, t \in R$, of $O(E)$ by taking infinitesimal generators $\alpha = \frac{d}{dt}g_t|_{t=0}$. With these notations we establish

Theorem 10.2 *Let $X(C)$ be given by (10.5.1) with $X(s)$ in $\mathcal{H}_1$, and assume that C runs only through $\mathbf{C}_0$. Then, the variation $\delta X(C)$ of $X(C)$*

is expressed in the form

$$\delta X(C) = \sum_j dt_j \int_C \{\alpha_j^C((FX)(s)\delta_j(s)dv(s) + (FX)(s)\delta_j(dv(s))\}, \tag{10.5.2}$$

where α_j^C *is the component of* α_j *normal to* C *and where* $\delta_j(s)$ *denotes the difference between* C *and* $C + \delta C$, *and where* $\delta_j(dv(s))$ *stands for the infinitesimal difference of the surface element* dv *at* s.

Proof. First apply $\mathcal{S}$-transform to the expression (10.5.1) so that we obtain an ordinary functional of ξ and C. Then, we appeal to the classical theory of calculus of variations (see e.g. Lévy [47]), where we see a formula for a functional $I = \int_C u ds$, C: contour in R^2,

$$\delta I = \int_C (\delta u \, ds + u\delta(ds)), \tag{10.5.3}$$

with the line element ds along the curve. The conclusion (10.5.2) can be proved by paraphrasing the above formula, and by extending the result to the case of higher dimensional manifold.

We then consider a white noise integral

$$X(x) = \int_{C_0} F(s)x(s)dv(s), \quad x \in E^*, \tag{10.5.4}$$

where C_0 passes through the origin. The diameter of C_0 is denoted as $\overline{oa}$.

Consider now the subgroup of $O(E)$ which leaves the C_0 invariant. Such a group, denoted by G_a, involves a subgroup isomorphic to the group generated by special conformal transformations, the isotropic dilation on R^{d-1} and the isotropy group at a, which is isomorphic to $SO(d-1)$.

Let H denote the Hilbert space $L^2(C_0, dv)$ and define U_g by

$$(U_g f)(v) = f(gv)|J|^{1/2}, \quad f \in H, \ J: \text{ Jacobian.} \tag{10.5.5}$$

Then, we can easily prove the following proposition by applying the reflection with respect to the unit sphere to show that the group G_a is isomorphic to the homothety group acting on R^{d-1}.

Proposition 10.4 *The unitary representation* $U : \{U_g; g \in G_a\}$ *of the group* G_a *on* H *is irreducible.*

Note that U is identified with a subgroup of $O(E)$.

Theorem 10.3 *Let $X(x)$ be defined by (10.5.4). Then, the space spanned by $\{X(g^*x); g \in G_a\}$ coincides with the space spanned by the system $\{\langle x, \xi\rangle : \xi \in C^\infty(C_a)\}$.*

Proof. Observe the expression of $X(x)$ in (10.5.4) and apply g^* to x. Then we have

$$X(g^*x) = \int_{C_a} (gF)(s)x(s)dv(s).$$

Since gF, $g \in G$, generates dense subset of $L^2(C_a, dv)$, $x(s)$ can be recovered, and the theorem is proved.

[2] Let **C** be the class of all possible C^∞-manifolds isomorphic to a sphere. Consider the random fields with parameter set **C** that are very much restricted.

Theorem 10.4 *Let $X(u, C)$ be the field given by*

$$X(u, C) = \int_D G(u, v; C)x(v)d\sigma(v). \tag{10.5.6}$$

Where $G(u, v; C)$ is Green's function. Then, we have

$$\delta X(u, C) = \int_D \delta G(u, v; C)x(v)d\sigma(v). \tag{10.5.7}$$

Proof. The $\mathcal{S}$-transform of the random variable $X(u, C)$ is given by

$$\{\mathcal{S}X(u, C)\}(\xi) = \int_D G(u, v; C)\xi(v)d\sigma(v).$$

Take its variation when C changes by δC. Then, we have

$$\int_D \delta G(u, v; C)\xi(v)d\sigma(v).$$

Applying the $\mathcal{S}^{-1}$-transform, we obtain (10.5.7).

Remark 10.2 *The first and the second terms of the right hand side of (10.5.7) can be discriminated, since they have different order in the mean square. Note that δG is given explicitly by the Hadamard equation (see Section 1.2 and [15]).*

10.6 Mathematical biology

There is another application, randomization of the Lotka–Volterra equation in population biology. The attempt towards this direction is wide and numerous, however we take a particular case, still giving us a suggestion.

Consider a simple example and start with a Lagrangian form which is randomized. With the notation established by Volterra, set $N_r(t)$ be the sight of the population r and $N_r(t) = \int_0^t N_r(s)ds$. Define

$$\chi = \sum_r \beta_r X_r' \log X_r', \quad 1 \le r \le n, \tag{10.6.1}$$

the demographic potential

$$P = \sum_r \beta_r \epsilon_r X_r + \frac{1}{2} \sum_{r,s} c_{r,s} X_r X_s, \tag{10.6.2}$$

and a bilinear form

$$Z = \sum_{r,s} a_{r,s} X_r' X_s. \tag{10.6.3}$$

Then, the Lagrangian form L, which is a functional of X_r's, is given by

$$L = \chi + \frac{1}{2} Z + P. \tag{10.6.4}$$

By the usual manner of the variational calculus to find the stationary point of the demographic action U:

$$U = \int_0^t L\,dt, \tag{10.6.5}$$

the Euler equation is given:

$$\frac{d}{dt}\frac{\partial L}{\partial X_r'} - \frac{\partial}{\partial X_r} = 0. \tag{10.6.6}$$

Thus we have

$$\frac{dN_r}{dt} = \left(\epsilon + \frac{1}{\beta_r} \sum_s a_{s,r} N_s \right) N_r, \tag{10.6.7}$$

where $N_r = X_r'$. Now we come to a random environment, namely the coefficients of the linear term of the demographic potential is taken to be a random variable. A good example is introduced by replacing ϵ_r with

$\epsilon_r + \dot{B}_r(t)$. Namely,

$$\frac{dN_r}{dt} = f_r(N) + \partial_t^* N_r, \tag{10.6.8}$$

where $(\epsilon + \frac{1}{\beta_r}\sum_s a_{s,r}N_s)N_r$ is simply written by $f_r(N)$.

The solution to this equation can be obtained as is done in [M. F. Dimentberg. Phys. Rev. E65 (2002), 1–6], however we only note the significance of this direction, specifically noting the idea to randomize the Lagrangian form.

We can see many other application in Quantum dynamics, Statistical mechanics, Game theory (random case), Maximum entropy methods, mathematical biology and etc.

10.7 Tomonaga–Schwinger equation

We start with the Schrödinger equation (Monin–Yaglom [58] Vol. 2 Section 28)

$$i\frac{\partial}{\partial t}\psi(t) = \mathcal{H}\psi(t),$$

where $\mathcal{H}$ is a Hamiltonian consisting of free part $\mathcal{H}_f$ and interaction part $\mathcal{H}_{\text{int}}$. Let $\mathcal{H}$ be expressed in the form

$$\begin{aligned}\mathcal{H} &= \int \mathcal{H}(u)du^3\\ &= \int \mathcal{H}_f(u)du^3 + \int \mathcal{H}_{\text{int}}(u)du^3\\ &= \mathcal{H}_f + \mathcal{H}_{\text{int}}.\end{aligned}$$

Introduce a unitary operator $U(t)$ such that

$$U(t) = \exp[it\mathcal{H}_f].$$

The collection $\{U(t)\}$ forms a one-dimensional parameter unitary group. With this, we claim that free field is governed by the quantization which makes it to be relativistically invariant.

Now introduce a family of vectors

$$\Phi(t) = U(t)\psi(t).$$

Then it satisfies

$$i\frac{d}{dt}\Phi(t) = \left(U(t)\int \mathcal{H}_{\text{int}}(u)U(t)^{-1}du^3\right)\Phi(t).$$

Consider a basic case where $\mathcal{H}_{\text{int}}$ is expressed as a polynomial in field operators $\varphi_\alpha(u,t), \alpha \in A$, such that $\varphi_\alpha(u,t) = U(t)\varphi_\alpha(u)U(t)^{-1}$. Then we have

$$\begin{aligned} U(t)\mathcal{H}_{\text{int}}(\varphi_\alpha(u), \alpha \in A)U(t)^{-1} &= \mathcal{H}_{\text{int}}(U(t)\varphi_\alpha(u)U(t)^{-1}, \alpha \in A) \\ &= \mathcal{H}_{\text{int}}(\varphi_\alpha(u,t), \alpha \in A). \end{aligned}$$

Hence, we have

$$i\frac{d}{dt}\Phi(t) = \int \mathcal{H}_{\text{int}}(\varphi_\alpha(u,t), \alpha \in A)du^3\Phi(t).$$

Note that it is not clear that the above formula is Lorentz invariant. We shall consider this property in what follows. Now t is generalized to a surface C. Let C be a space-like hyper-surface in the 4-dimensional Minkovski space. We can consider a space vector $\Phi(C)$. Then we should have

$$\Phi(C + \delta C) = (1 - id\omega\mathcal{H}_{\text{int}}(\varphi_\alpha(u,t), \alpha \in A))\Phi(C),$$

where $d\omega = du^3dt$. Remind that $d\omega$ is Lorentz invariant. Hence, the above variational equation is independent of the choice of the coordinate system.

Now, the functional derivative can be defined

$$i\frac{\delta\Phi(C)}{\delta n(s)} = \mathcal{H}_{\text{int}}(s)\Phi(C),$$

where $\{\delta n(s)\}$ denotes the distance between C and $C + \delta C$. Note that $\delta n(s)$ may be written as δC_s symbolically, as was done in Example 7.4. Thus, we have obtained the Tomonaga–Schwinger equation.

We have so far proceeded in the relaxed view of mathematical rigor to come to the goal quickly. We have hope that an approach to the Tomonaga–Schwinger equation within the framework of white noise analysis would be fruitful and quite interesting, where the field operators are expressed in terms of operators in quantum white noise.

An additional note is that an integrability condition of the Tomonaga–Schwinger equation is given by

$$\frac{\delta}{\delta n(u)}\frac{\delta}{\delta n(v)}\Phi(C) = \frac{\delta}{\delta n(v)}\frac{\delta}{\delta n(u)}\Phi(C).$$

That is, for every u, v in space-like hypersurface C, we have

$$\mathcal{H}_{\text{int}}(v)\mathcal{H}_{\text{int}}(u) = \mathcal{H}_{\text{int}}(u)\mathcal{H}_{\text{int}}(v).$$

Appendix

A.1 Appendix 1

Characteristic functionals

The roles of characteristic function of a probability distribution or of a random variable is quite well-known. An infinite dimensional analogue is the characteristic functional, which corresponds to a probability distribution of a stochastic process. In this case, a particular attention should be paid to the probability distributions of *generalized stochastic processes*, for which probability distributions are introduced on the space of generalized functions. An easy generalization of a characteristic function, which corresponds to a probability distribution on a finite dimensional space, cannot be established in the infinite dimensional case, like a function space.

A quick review of a characteristic function is done as follows. Let a probability distribution m on R^d be given. Then, the characteristic function $\varphi(t)$ is given by

$$\varphi(t) = \int_{R^d} \exp[i(x,t)]dm(x), \quad t \in R^d, \tag{A.1.1}$$

where $(\cdot,\cdot)$ is the inner product in R^d.

The $\varphi(t)$ satisfies

(1) $\varphi(z)$ is a continuous function of z,
(2) $\varphi(0) = 1$,
(3) Positive definite: For any $t_j, 1 \le j \le n$, and for any complex numbers $z_j, 1 \le j \le n$, the following inequality holds.

$$\sum_1^n z_j \bar{z}_k \varphi(t_j - t_k) \ge 0. \tag{A.1.2}$$

Conversely, we are given $\varphi(t)$ satisfying $(1), (2), (3)$, then there exists a probability distribution m on R^d such that the relation (A.1.1) holds.

Roughly speaking, there is a one-to-one correspondence between a probability distribution and a characteristic function. (S. Bochner's Theorem). For actual computations of probability distributions and for the determination of distributions, the use of characteristic function is very efficient.

We now come to an infinite dimensional analogue, although not quite an analogue. There is an important theorem due to R.A. Minlos, we call it the Bochner–Minlos Theorem. (For details, see [18].)

Let E, a dense subspace of $L^2(R^d)$, be a countably Hilbert nuclear space, topologised by countably many compatible Hilbertian norms $\|\cdot\|_n, n \geq 0$, with respect to which E is complete. Let E_n be the completion with respect to the n-th norm $\|\cdot\|_n$, which are arranged in increasing order. Then we have

$$E = \bigcap_n E_n.$$

Letting $\|\cdot\|_{-n}$ be the norm of the Hilbert space E_n^*, the dual space of E_n. We see that $\{\|\cdot\|_n : -\infty < n < \infty\}$ is an increasing family of Hilbertian norms. The dual space E^* of E can be expressed as

$$E^* = \bigcup_n E_n^*.$$

Definition A.1 Let E be the space given above. If for any m there exists $n > m$ such that the injection mapping

$$T_m^n : E_n \to E_m$$

is nuclear, then E is called a *countably Hilbert nuclear space* or simply a *nuclear space.*

Now the basic Hilbert space is $L^2(R^d)$ and the canonical bilinear form that links E and E^* is denoted by $\langle\cdot,\cdot\rangle$ to discriminate from the inner product in $L^2(R^d)$.

Definition A.2 A *characteristic functional* is a functional $C(\xi)$ defined on E such that

(1) $C(\xi)$ is a continuous functional of $\xi \in E$,
(2) $C(0) = 1$,

(3) $C(\xi)$ is positive definite, i.e. for any $\xi_j, 1 \le j \le n$, and for any complex numbers $z_j, 1 \le j \le n$, the following inequality holds.

$$\sum_1^n z_j \bar{z}_k C(\xi_j - \xi_k) \ge 0. \tag{A.1.3}$$

Then, we have the theorem

Theorem A.1 *(Bochner–Minlos) Given a characteristic functional $C(\xi)$ on E, then there exists a probability measure μ on the measurable space $(E^*, \mathbf{B})$ such that*

$$C(\xi) = \int_{E^*} \exp[i\langle x, \xi\rangle] d\mu(x), \tag{A.1.4}$$

where $\mathbf{B}$ is the sigma-field generated by the cylinder subsets of E^*. In addition, such a measure μ is unique.

For proof, the reader is recommended to refer to Gel'fand–Vilenkin [12] or Hida [18].

Theorem A.2 *Let $C(\xi), \xi \in E$, be a characteristic functional. If*

(1) *$C(\xi)$ is continuous in $\|\cdot\|_m$ for some m,*
(2) *for some $n(>m)$ the injection $T_m^n : E_n \to E_m$ is of Hilbert–Schmidt type, then there exists a unique countably additive measure μ on $(E^*, \mathcal{B})$ and μ is supported by E_n^*.*

We omit the proof, but some examples are shown to get some idea.

Example A.1 White noise.

Let $C(\xi)$ be given by

$$C(\xi) = \exp\left[-\frac{\sigma^2}{2}\|\xi\|^2\right].$$

The measure μ_{σ^2} on $(E^*, \mathbf{B})$ determined by the above characteristic functional is called a *Gaussian white noise* with variance σ^2. If σ is taken to be 1, then we call it *standard Gaussian* white noise, or simply *white noise* and denoted by μ, which is the white noise measure.

The measure μ_{σ^2} is *invariant* under the infinite dimensional rotation group ($O(E^*)$-invariant) and even *ergodic*.

To be concrete, the measure μ is viewed as the probability distribution of the time derivative of a Brownian motion $B(t)$, also called a white noise and denoted by $\dot{B}(t)$. Thus, μ-almost all x in E^* is a sample function of $\dot{B}(t)$.

Example A.2 Poisson noise.

The characteristic functional

$$C_P(\xi) = \exp\left[\lambda \int (e^{i\xi(t)} - 1)dt\right], \quad \lambda > 0,$$

determines *Poisson noise* μ_P. The parameter λ is the intensity of the Poisson noise. The probability distribution of the derivative of a Poisson process is given by this μ_P.

Example A.3 Compound Poisson noise.

The characteristic functional $C_{cP}(\xi)$ of this noise is of the form

$$C_{cP}(\xi) = \exp\left[\lambda \int_R \left\{\exp[i\xi(t)u] - 1 - i\xi(t)\frac{u}{1+u^2}\right\} du\right],$$

where $dn(u)$ is the Lévy measure such that

$$\int_{R-\{0\}} \frac{u^2}{1+u^2} dn(u) < \infty.$$

This functional defines the probability distribution of a generalized stochastic process obtained by the time derivative of a compound Poisson process, which is a Lévy process without Gaussian part.

In general, the time derivative of a Lévy process $Z(t)$ has a characteristic functional of the form

$$C_Z(\xi) = C_G(\xi)C_{cP}(\xi).$$

The $C_G(\xi)$ in the above formula is of the form

$$C_G(\xi) = \exp\left[-\frac{\sigma^2}{2}||\xi||^2\right].$$

This is a general case corresponds to a system of i.e.r.v.'s (see Section 2.1).

So far we have discussed the probability distribution of generalized stochastic processes with *independent values at every (space-time) point.* In fact, the characteristic functional satisfies a functional equation

$$C(\xi_1 + \xi_2) = C(\xi_1)C(\xi_2),$$

whenever $\xi_1(u) \cdot \xi_2(t) = 0$.

Example A.4 Wiener measure.

The probability distribution of a Brownian motion $\{B(t); t \geq 0\}$ is called the Wiener measure. Let it be denoted by μ^W. It has the characteristic functional of the form

$$C(\xi) = \exp\left[-\frac{1}{2}\int_0^\infty |\tilde{\xi}(u)|^2 du\right],$$

where $\tilde{\xi}(t) = \int_t^\infty \xi(u)du, \quad t \geq 0$.

By the Kolmogorov–Prokholov theorem (see e.g. [18]) it is proved that μ^W is supported by the space $C[0, \infty)$ of continuous functions on $[0, \infty)$.

Here is a short note regarding the data processing of observed data, then one can obtain estimates of some basic statistics, like the mean, the covariance or the spectral density.

In principle, the characteristic functional gives us full information on the probability distribution of a stochastic process $X(t)$. It is therefore recommended to obtain the characteristic functional from the data. It is, by definition, the expectation of $\exp[i \int X(t)\xi(t)dt]$ over all possible trajectories $X(t, \omega), \omega \in \Omega$. In fact, it is possible to take the set $\{X(\cdot, \omega), \omega \in \Omega(P)\}$, but we can appeal to the ergodic property of the shift $\{S_t\}$, provided that $X(t)$ is a functional of white noise and that the transition from $X(t)$ to $X(t+h)$ comes from $U_h X(t) = X(t+h)$, where the unitary operator is defined by S_t as in Section 10.3.

In short, $X(t)$ is the output signal obtained by white noise input. With this assumption, the actual computation of the characteristic functional $C_X(\xi)$ is as follows:

$$C_X(\xi) = \lim_{T \to \infty} \frac{1}{T} \int_a^{T+a} e^{i\langle U_t X(0), \xi\rangle} dt. \tag{A.1.5}$$

Under some weaken assumption, for example, the spectral density is flat, i.e. constant, the ergodic hypothesis still holds.

A.2 Appendix 2

A review of the variation of functional $U(f)$

A short review of the classical theory of calculus of variations for a complex valued functional $U(f)$, where f runs through a certain functional space, say a Fréchet space F. Such a functional is interesting in itself, and also it presents a representation of a functional $U(C)$, C being some manifold, although such a representation is not unique.

Suppose $U(f)$ is continuous. If, for an infinitesimal variation δf of $f \in F$, the following conditions

(1) there exists a functional $\delta U(f, \delta f)$ such that

$$U(f + \delta f) - U(f) = \delta U(f, \delta f) + o(\delta f),$$

(2) $\delta U(f, \delta f)$ is continuous linear in δf, holds then, the $\delta U(f, \delta f)$ is called a *variation* of $U(f)$.

If the variation of $U(f)$ exists, then by the condition (2), there is a function $U'_f(f, t)$ of t depending on f such that

$$\delta U(f, \delta f) = \langle U'_f(f, \cdot), \delta f \rangle,$$

where the angular bracket is the canonical bilinear form connecting F and its dual space F^*.

Definition The functional $U'_f(f, t)$ is called the functional derivative (of the first order) of $U(f)$.

We then come to the second order functional derivatives.

Suppose the variation of $\delta U'_f(f, t)$ of the functional derivative can formally be expressed in the form

$$\delta U'_f(f, t) = \langle U''_{ff}(f) + U''_{fg}(f), \delta f \rangle,$$

where

$$\langle U''_{ff}(f), \delta f \rangle = U''_{ff}(t) \delta f(t)$$

and where

$$\langle U''_{fg}(f), \delta f \rangle = \int U''_{fg}(t, s), \delta f(s) ds.$$

Thus, there are two functional derivatives of order two; namely $U''_{ff}(t)$ and $U''_{fg}(t, s)$.

Formally writing, the second order variation $\delta^2 U$ of U in this case may be expressed in the form

$$\delta^2 U = \int U''_{ff}(t)[\delta f(t)]^2 dt + \int \int U''_{fg}(t, s) \delta f(t) \delta f(s) dt\, ds.$$

Because of this, we call U''_{ff} and U''_{fg} the *singular part* and the *regular part* of the second order functional derivative, respectively.

A review of the calculus for non-random fields $G(C)$

We first prepare some background of the variational calculus for non-random real-valued function $G(C)$ defined on a topological space $\mathbf{C}$, which is given by

$$\mathbf{C} = \{C;\ \ C \approx S^{d-1},\ \text{convex}\},$$

where $\approx$ means smooth homeomorphism. The topology is introduced to $\mathbf{C}$ by using the Euclidean metric.

Consider an infinitesimal deformation of C within $\mathbf{C}$, which is symbolically denoted by $C + \delta C$. To concretize such a deformation, we need an analytic expression of the amount of the deformation δC:

$$\delta C \simeq \{\delta n(s);\ \ s \in C\},$$

where $\delta n(s)$ represents the outward normal vector to C at the point $s \in C$. We assume that the $\delta n(s)$ is a smooth function of s, say in E. The topology introduced on $\mathbf{C}$ can be defined in such a way that $C + \delta C$ converges to C, if and only if $\|\delta n\| \to 0$, where $\|\delta n\| = \sup_s\{|\delta n(s)| + |\delta n'(s)|\}$.

Take a real-valued, non-random functional $G(C)$ of C, and assume that the $G(C)$ satisfies the following:

$$G(C + \delta C) - G(C) = \delta G(C) + g(C, \delta C)$$

with the conditions

(i) $\delta G(C)$ is continuous and linear in $\delta n(s)$, and
(ii) $g(C, \delta C)$ is of order $o(\|\delta n\|)$.

The $\delta G(C)$ satisfying these conditions is called the *variation* of $G(C)$ at C.

The fact (i) implies that there is a function $\varphi(s)$ such that the variation $\delta G(C)$ can be expressed in the form

$$\delta G(C) = \int_C \varphi(s)\delta n(s)ds.$$

If $\varphi(s)$ is continuous, then, we denote $\varphi(s)$ by $\frac{\partial G(C)}{\partial n}(s)$ and call it the *functional derivative* of $G(C)$.

In this case, one may write

$$\delta G(C) = \int_C \frac{\partial G(C)}{\partial n}(s)\delta n(s)ds.$$

Remark A.1 *From now on, the δn is positive if it is measured outward unless contrary is noted. In fact, the interior of the domain C, enclosed* *by C is tacitly understood to be the past, in a sense, so that δC is taken towards the future (in the positive direction). (See Fig. 7.)*

The *second order variation* of $G(C)$ is defined by $\delta(\delta G(C))(= \delta^2 G(C))$, and the second order functional derivative can be defined by the second order variation of $G(C)$. Let $\delta^2 G(C)$ be expressible as

$$\delta^2 G(C) = \int_{C \times C} \psi(t,s)\eta \otimes \eta(t,s) dt\, ds + o(\eta \otimes \eta).$$

In general ψ is a generalized function, however if it is a continuous function, then it is called the second order functional derivative and denoted by $\frac{\partial^2 G}{\partial \eta \partial \eta}(s,t)$. It is a symmetric function. A favourable and indeed somewhat interesting case is that it is expressed as a sum of two terms as follows:

$$f(s)\delta(t-s) + g(s,t),$$

where f is continuous and g is an L^2-kernel. Note that both are functional of ξ. Customary, the two functions are denoted by $G''_{\xi^2}(s)$ and $G''_{\xi,\xi'}(s,t)$, respectively.

As we shall see later, the integral of the first term

$$\Delta_L G = \int G''_{\xi^2}(s) ds$$

determines the Lévy Laplacian Δ_L in functional analysis, acting on non-random functional $G(C)$. It should be noted that for the first term the singularity is found only on the diagonal. This fact suggest a special role in the analysis of generalized white noise functionals discussed in later sections.

The variational calculus for a (non-random) function of C is more efficiently used for the calculus of a random field rather than a function $\Phi(f)$, f being a function. Note that C has an analytic representation in terms of a function, for which the classical theory of functionals can be applied. There we should note the analysis does not depend on the choice of a function f.

Sometime, the Lévy Laplacian is defined by the average over an interval T,

$$\frac{1}{|T|} \int_T G''_{\xi^2}(s) ds,$$

$|T|$ being the length of T, if the parameter set is limited.

A.3 Appendix 3

Reproducing kernel Hilbert space (RKHS)

General theory

Let F be a class of functions defined on E, forming a Hilbert space (real or complex) with an inner product $(\cdot,\cdot)$. A kernel $K(x,y),(x,y) \in E \times E$, is called a *reproducing kernel* of F if

(1) For every y, $K(x,y)$, as a function of x, belongs to F
(2) the reproducing property holds : for every $y \in E$ and every $f \in F$

$$(f(x), K(x,y)) = f(y).$$

The Hilbert space F with reproducing kernel K is called a *reproducing kernel Hilbert space.* It is often written as $H(K)$.

By definition the following properties are immediately verified.

1. If a reproducing kernel exists, then it is unique.
2. For any fixed $y \in E$, the function $y(f) \equiv f(y)$ is a continuous functional of $f \in F$.
3. The reproducing kernel K is positive definite, more precisely non-negative definite: for any n, any $y_j, 1 \leq j \leq n$ and any ξ_j in $\mathbf{C}$, $1 \leq j \leq n$,

$$\sum_{i,j} K(y_i, y_j)\xi_i\bar{\xi}_j \geq 0.$$

In particular

$$K(x,x) \geq 0, \quad K(x,y) = \overline{K(y,x)}, \quad |K(x,y)|^2 \leq K(x,x)K(y,y).$$

4. If f_n strongly converges to f in the space $H(K)$, then f_n converges to f pointwise. In addition, if $K(x,x)$ is uniformly bounded, then the converges of f_n is uniform.
5. If $H(K)$ is a subspace of a bigger Hilbert space F, then the equation

$$f(y) = (h(x), K(x,y)), \quad h \in F$$

gives the projection f of the element h in F.

6. If $\{g_n\}$ is a complete orthonormal system in $H(K)$, then for every sequence $\{\alpha_n\}$ with $\|\alpha\|^2 = \sum_n |\alpha_n|^2 < \infty$, we have

$$\sum_1^\infty |\alpha_n||g_n(x)| \leq K(x,x)\frac{1}{2}\|\alpha\|$$

and

$$K(x,y) = \sum_n g_n(x)g_n(y).$$

Remark A.2 *Property 4 is quite helpful for us when white noise analysis is discussed.*

Construction
Let D be an abstract set. Given a positive definite kernel $K(x,y)$ on $D \times D$. Then, the following theorem is proved.

Theorem A.3 *There is a complex Hilbert space* $\mathbf{H}(K)$ *of functions on* D *such that*

(1) *the system* $K(\cdot,t), t \in D$ *spans the entire space* $\mathbf{H}(K)$
(2) *the inner product in* $\mathbf{H}(K)$ *satisfies*

$$(f(\cdot), K(\cdot,t)) = f(t),$$

for any f *in* $\mathbf{H}(K)$.

Namely, we can construct a reproducing kernel Hilbert space $H(K)$, for which the given kernel $K(x,y)$ is the reproducing kernel.

Outline of the proof
(i) Let F_1 be a complex vector space spanned by the vectors of the form

$$f_1(x) = \sum_{k=1}^{n} a_k K(x, y_k), \quad a_k \in \mathbf{C}.$$

Define a bilinear form $(f_1, g_1), f_1, g_1 \in F$, by

$$(f_1, g_1) = \sum_{k=1}^{n} \sum_{j=1}^{m} a_k \bar{b}_j K(u_j, y_k)$$

with

$$g_1(x) = \sum_{j=1}^{m} b_j K(x, u_j).$$

(ii) Introduce an equivalence relation $\sim$ to the space F, by

$$f_1 \sim f_1' \Leftrightarrow \|f_1 - f_1'\| = 0.$$

Then, F_1 is classified to obtain $F = F_1/\sim$. A member of F is denoted by $f, g, \ldots$ It is easy to see that F is a vector space to which the inner product (f, g), is introduced by using (f_1, g_1).

Define $\|f\|$ for $f \in F$ by $\|f\| = \|f_1\|, f_1 \in F_1$, f being the representative of the class f. The $\|\ \|$ is a Hilbertian norm $\|\ \|$. For a fixed $y \in D$, set $g(x) = K(x, y)$. Then for $f_1(x) = \sum_{k=1}^{n} a_n K(x, y_k)$, we have

$$(f(\cdot), K(\cdot, y)) = \left(\sum_{k=1}^{n} a_k K(\cdot, y_k), K(\cdot, y)\right) = \sum_{k=1}^{n} a_k K(y, y_n) = f(y),$$

which shows that $K(x, y)$ is a reproducing kernel.
(iii) If necessary, we take the completion of the normed space F to have a Hilbert space $H(K)$. The kernel $K(x, y)$ plays the role of the reproducing kernel on $H(K)$.

If the diagonal values of the kernel $K(x, y)$ is bounded, that is $K(x, x) \leq M$ for every x and for some M, then the reproducing property (2) proves that $|f| \leq M\|f\|$ for every $f \in \mathbf{H}(K)$. Then follows that the weak convergence in $\mathbf{H}(K)$ implies the point-wise convergence.

These facts are the case where the kernel comes from a characteristic functional of a stochastic process.

As we have seen in Section 2.1, a characteristic functional $C(\xi)$ is positive definite on the space E, so that it defines a reproducing kernel Hilbert space with the kernel $C(\xi - \eta), (\xi, \eta) \in E \times E$.

Example A.5 White noise.

The characteristic functional is $C(\xi) = \exp[-\frac{1}{2}\|\xi\|^2]$. Take $C(\xi - \eta)$ to be the reproducing kernel to define the RKHS $\mathbf{H}(C)$.

Write $C(\xi - \eta)$ in the form

$$C(\xi - \eta) = C(\xi) \exp[\langle \xi, \eta \rangle] c(\eta).$$

Then, we have

$$\frac{d}{dt} C(\xi - t\eta)|_{t=0} = C(\xi)\langle \xi, \eta \rangle.$$

By repeating this procedure, we obtain a functional of ξ of the form

$$C(\xi) H_n(\langle \cdot, \eta \rangle; \|\eta\|^2),$$

where $H_n(u, \sigma^2)$ denotes the Hermite polynomial of degree n with parameter σ^2. We can now recognize that $H_n(\langle \cdot, \eta \rangle; \|\eta\|^2)$ appears as the S-transform of a monomial $\langle x, \xi \rangle^n$ of degree n.

Let the above trick be generalized to the case where η and the differential operator $\frac{d^k}{dt^k}$ are replaced by $\sum t_j \eta_j$ and $\Pi_k \left(\frac{\partial}{\partial t_j}\right)^{k_j}$, respectively. The η_j's are taken to be orthonormal.

Then, we can prove

Theorem A.4 *The RKHS* $\mathbf{H}(C)$ *defined by the kernel* $C(\xi - \eta), (\xi, \eta) \in E \times E$ *in the case of white noise is isomorphic to the function space* $\mathbf{F}$ *given by the* S*-transform in such a way that a homogeneous chaos in* (L^2) *corresponds to a monomial on* ξ *up to a factor* $C(\xi)$*.*

A similar argument can be applied to the case of Poisson noise, by replacing monomial in ξ with a monomial in $e^{i\xi}$.

Example A.6 Gaussian process.

Let $X(t), t \in I = [0,1]$ be a mean continuous Gaussian process. We assume $E[X(t)] = 0$. Let $\Gamma(t,s)$ be the covariance function of $X(t)$. Since $\Gamma(t,s)$ is positive definite, there is a RKHS $H(P)$. It is isomorphic to the Hilbert space $M(X)$ spanned by the $X(s), s \in I$. The isomorphism is determined by the mapping Π:

$$\Pi : X(t) \leftarrow \Gamma(\cdot, t) \in H(P).$$

Since Γ is bounded, the mean convergence turns into a point-wise convergence in $H(\Gamma)$.

In addition, the subspace $M_t(X)$ spanned by $X(s), s \le t$, is mapped under Π to a subspace $H_t(\Gamma)$ of $H(\Gamma)$:

$$\Pi(M_t(X)) = H_t(\Gamma).$$

Note that $H_t(\Gamma)$ is spanned by the $\Gamma(u,s), s \le t, u \in I$.

By the property 5 in general theory, the prediction can be discussed by using the subspace $H_t(P)$.

Example A.7 Gaussian random field $X(C), C \in \mathbf{C}$.

Assume that $X(C)$ has a canonical representation in term of white noise $x(u), u \in R^d$. Let $X(C)$ be represented by

$$X(C) = \int_C F(C,u)x(u)du.$$

Let $\Gamma(C, C')$ be the covariance function of $X(C)$. Then, there is a RKHS with reproducing kernel $\Gamma(C, C')$. There a similar game to Example A.6 can be played.

A.4 Appendix 4

Poisson–Charlier polynomials

For Poisson noise the so called Poisson–Charlier polynomial plays the role of the Hermite polynomial for Gaussian white noise in the analysis of the (L^2)-space.

Let $p(x;\lambda)$ be the Poisson distribution with parameter $\lambda > 0$:

$$p(x;\lambda) = \frac{\lambda^x}{x!}e^{-\lambda}, \quad x = 0, 1, 2, \ldots$$

The Poisson–Charlier polynomial $p_n(x;\lambda)$ is given by

$$p_n(x;\lambda) = (-1)^n \frac{\lambda^{n/2}}{\sqrt{n!}} \frac{\Delta^n p(x;\lambda)}{p(x;\lambda)},$$

where Δ is the difference operator:

$$\Delta f(x) = f(x) - f(x-1)$$

and where $p(x;\lambda) = \frac{\lambda^x}{x!}e^{-\lambda}$, Poisson distribution.

The polynomials $p_n(x;\lambda), n = 0, 1, 2, \ldots$ form an orthonormal system in the following sense:

$$\sum_{x=0}^{\infty} p_n(x;\lambda) p_m(x;\lambda) p(x;\lambda) = \delta_{n,m}.$$

Examples

$$\begin{aligned}
p_0(x,\lambda) &= 1 \\
p_1(x,\lambda) &= \frac{1}{\sqrt{\lambda}}(x - \lambda) \\
p_2(x,\lambda) &= \frac{1}{\sqrt{2}\lambda}(x^2 - (1+2\lambda)x + \lambda^2) \\
p_3(x,\lambda) &= \frac{1}{\sqrt{6}\lambda^{3/2}}\left(x^3 - 3(1+\lambda)x^2 + (3\lambda + 3\lambda^2 + 2)x - \lambda^3\right) \\
\cdots\cdots &\quad \cdots\cdots\cdots
\end{aligned}$$

These polynomials form a complete orthonormal system with respect to the density $\frac{\lambda^x}{x!}e^{-\lambda}$.

A relation to the Laguerre polynomial $L_n^\alpha(a)$

The equation (6.3.2) can be obtained from

$$L_n^{\alpha+\beta+1}(a+b) = \sum_{m=0}^{n} L_m^\alpha(a) L_{n-m}^\beta(b)$$

and

$$L_n^{x-n}(a) = \frac{a^{n/2}}{\sqrt{n!}} p_n(x,a).$$

Epilogue

(I) The Chern–Simons action integrals

At the last stage of writing this volume, we knew that our white noise analysis, in particular the theory of generalized white noise functionals (white noise distributions), will play important roles in the path integral theory of Chern–Simons action functionals too. Development towards this direction seems to be quite interesting. It is, however, impossible to explain details on how it will work. We now wish to give just a very short note to show what we wish to discuss on this topic.

Let G be a connected Lie group and let $\mathbf{g}$ be the Lie algebra with an Ad-invariant inner product $\langle \cdot, \cdot \rangle_{\mathbf{g}}$. A connection A is a smooth $\mathbf{g}$-valued 1-form on $R^3 = \{(x_0, x_1, x_2); x_i \in R\}$.

The Chern–Simons action CS is given, with the usual notation, by

$$CS(A) = \frac{\kappa}{4\pi} \int_{R^3} \left(\langle A \wedge dA \rangle + \frac{1}{3} \langle A \wedge [A, A] \rangle \right) dv, \tag{1}$$

where dv is the Lebesgue measure on R^3.

Assume the gauge invariance of $\exp[iCS(A)]$ and others. Then, one may consider the case where A is expressed in the form

$$A = a_0 dx_0 + a_1 dx_1,$$

where a_0 and a_1 are $\mathbf{g}$-valued smooth function on R^3 with the boundary conditions:

$$a_1(x_0, x_1, 0) = 0, \quad a_0(x_0, 0, 0) = 0. \tag{2}$$

Now $CS(A)$ given by (1) is written as $CS(a_0, a_1)$ and we can prove

$$CS(a_0, a_1) = \frac{\kappa}{2\pi} \langle a_0, -\partial_2 a_1 \rangle = \frac{\kappa}{2\pi} \langle a_0, -f_1 \rangle,$$

with $f_1 = \partial_2 a_1$.

The Chern–Simons integral is of the form

$$\int_{\mathcal{A}'} \phi(a_0, a_1) e^{iCS(a_0,a_1)} Da_0 Df_1, \tag{3}$$

where $\mathcal{A}'$ is the collection of A's satisfying the conditions (2).

Now follow interesting problems.

(1) The measures Da_0 and Df_1 may basically be assumed to be Gaussian. Taking the condition (2) into account, one may think of some modification. Then, come to the path integrals.

(2) Lagrangian dynamics for $CS(A)$ has been discussed successfully to some extent. One can expect more development.

(3) Since G is taken to be a gauge group, can one consider complex white noise and infinite dimensional unitary group?

The authors have received new insights from the works such as

P. Leukert and J. Schaefer, Reviews in Math. Phys. 8 (1996), 445–456,

S. Albeverio and A. Sengupta, Commun. Math. Phys. 186 (1997), 563–579,

and recent paper [43] listed in the Bibliography.

(II) The elemental Poisson noise

From the view point of Reductionism we understand that Gaussian white noise and Poisson noise are elemental. Also, we understand that, in an intuitive level, innovation can be thought of a composition of Gaussian white noise and superposition of Poisson noises with different jumps under certain reasonable assumptions. This should be proved in a level of mathematical rigor.

As the reader can guess, this fact needs profound discussion to be well understood. From a certain point of view, we can prove by taking a generalized random field with two-dimensional parameter (t, u), where $t \geq 0$ stands for the time and $u \in R - \{0\}$ tacitly corresponds to the height of jumps of a Poisson process (noise) which is an elemental component of the compound Poisson noise. A candidate of the generalized random field in question, denote it by $\dot{P}(t, u)$, has the characteristic functional of the following form

$$C(\xi) = \exp\left[\lambda \int\int (e^{i\xi(t,u)} - 1) dt\, du\right],$$

where ξ is a test functional in a certain nuclear space.

Behind our analysis in this volume is the investigation of the $\dot{P}(t, u)$ and its nonlinear functionals. For rigorous discussion we refer to our forthcoming paper

T. Hida and Si Si, Note on the Lévy field (to appear).

List of Notations

$C(\xi), C_P(\xi), C_L(\xi), C_P(\xi),$		2.1, 3.1, 3.8
(L^2)		2.1
$(S) \subset (L^2) \subset (S)^*$		2.1
$(L^2)_p^-$		3.1
$: x^{\otimes n} :$	Wick product	2.1
$(L^2) = \bigoplus H_n$	Fock space	2.1
$O(E), O^*(E^*)$	infinite dimensional rotation group	2.3
$\Delta_\infty, \mathcal{N} = -\Delta_\infty$		2.4
$\mathcal{G}$	Lévy group	2.4
$\mathcal{H}$		2.4
Δ_L, Δ_V	Lévy and Volterra Laplacian	2.5
$(\mathbf{P})$		3.9
$\mathbf{C}$		4.5
$D_0(R^d)$		10.1
$\mathbf{H}(K)$	reproducing kernel Hilbert space	3.1, 11.3
$C(d), C(D)^*$	conformal grpoup	10.1
U''_{ff}, U''_{fg}		11.2

Bibliography

[1] L. Accardi *et al.*, eds., Selected papers of Takeyuki Hida, World Sci. Pub. Co. Ltd, 2001.

[2] L. Accardi, T. Hida and Si Si, Innovation approach to some stochastic processes and random fields, Univ. Roma II, Volterra Center.

[3] L. Accardi, Y.G. Lu and I. Volovich, Quantum Theory and Stochastic limits, Springer-Verlag, 2002.

[4] N.N. Chentsov, Lévy's Brownian motion for several parameters and generalized white noise, Theory of Prob. Appl. 2 (1957), 265–266.

[5] R. Courant and D. Hilbert, Methods of mathematical physics, vol. I. Wiley, 1953.

[6] R.C. Dalang and J.B. Walsh, The sharp Markov property of the Brownian sheet and related processes, Acta Math. 108 (1992), 153–218.

[7] P.A.M. Dirac, The Lagrangian in quantum mechanics, Phys. Z. USSR 3 (1933), 64–72.

[8] Dobrushin and D. Surgailis, On the innovation problem for Gaussian random fields, Z. Wahrscheinlichkeits thorie und verwandte Gebiete 49 (1979) 275–291.

[9] M.D. Donsker and J.L. Lions, Fréchet-Volterra variational equations, boundary value problems, and function space integrals, Acta Math. 108 (1962), 147–228.

[10] I.M. Gel'fand, Generalized random processes (in Russian), Doklady Akad.Nauk S.S.S.R. 100 (1955), 853–856.

[11] I.M. Gel'fand and S.V. Fomin, Variational calculus (Russian), Phys-Math. Lit. Moscow, 1961.

[12] I.M. Gel'fand and N.Ya. Vilenkin, Generalized functions. vol. 4. (English transl.), Academic Press, 1964.

[13] J. Glimm and A. Jaffe, Quantum Physics, A functional integral point of view, Springer-Verlag, 1981.

[14] J. Hadamard, Lecons sur le calcul des variations, Hermann, 1910.

[15] J. Hadamard, Sur le problème d'analyse relatif à l'équilibre des plaques élastiques encastrées, Mémoire couronne par l'Académie des Sciences (Prix Vaillant) Mém. Sav. Etrang. 33. 1907. Supplément: Existence de la fonction Γ pour le plan sectionne. Collected Works vol. 2, 515–641.

[16] R. Hanbury Brown, The intensity Interfrometer, its Application to Astronomy, Taylor & Francis Ltd. 1974.

[17] T. Hida, Canonical representations of Gaussian processes and their applications, Memoirs Coll. Sci., Univ. Kyoto-A33 (1960), 109–155.

[18] T. Hida, Brownian motion, Springer-Verlag. 1980 (Original Japanese edition, 1975).

[19] T. Hida, K.-S. Lee and Si Si, Multidimensional Parameter white noise and Gaussian random fields, Balakrishnan volume, 1987, 177–183.

[20] T. Hida, A note on generalized Gaussian random fields, J. Multivariate Anal. 27 (1988) 255–260.

[21] T. Hida, White noise and stochastic variational calculus for Gaussian random fields, Dynamics and Stochastic Processes, Lisbon, 1988, Lec. Notes in Physics 335, (eds. R. Lima *et al.*). 1988, 136–141.

[22] T. Hida, Si Si, Variational calculus for Gaussian random fields, Proc. IFIP Warsaw, Lec. Notes in Control and Information Sci. (ed. J. Zabczyk) #136, 1989, 86–97.

[23] T. Hida, White noise analysis and Gaussian random fields, Proc. the 24th Winter School of Theoretical Physics, Karpacz, Stochastic Methods in Math. and Phys. (eds. R. Gielerak and W. Karwowski), 1989, 277–289.

[24] T. Hida, Functionals of Brownian motion. Lectures in Appl. Math. and Informatics. ed. L. Ricciardi, Manchester Univ-Press, 1990, 286–329.

[25] T. Hida, White noise and random fields – old and new, Proc. Gaussian Random Fields (ed. T. Hida and K. Saito), 1991, Part 3. 1–10.

[26] T. Hida, Stochastic variational calculus, Stochastic Partial Differential Equations and Applications (ed. B.L. Rozovskii and R.B. Sowers), Springer-Verlag, 1992, 123–134.

[27] T. Hida, White noise and Gaussian random fields, Probability Theory (ed. Louis H.Y. Chen), Walter de Ruyter & Co., 1992, 83–90.

[28] T. Hida, H.-H. Kuo, J. Potthoff and L. Streit, White noise – An infinite dimensional calculus, Kluwer Academic Pub. 1993.

[29] T. Hida, Random fields as generalized white noise functionals, Acta Applicandae Mathematicae 35, 1994, 49–61.

[30] T. Hida and Si Si, Stochastic variational equations and innovations for random fields, Infinite Dimensional Harmonic Analysis, Transactions of a German-Japanese Symposium (eds. H. Heyer and T. Hirai), 1995, 86–93.

[31] T. Hida, A note on stochastic variational equations, Exploring Stochastic Laws, (Korolyuk volume) (eds. A.V. Skorohod and Yu.V. Borovskikh), 1995, 147–152.

[32] T. Hida, Random fields and quantum dynamics, Foundations of Physics, 27, no.11, Namiki volume, 1997, 1511–1518.

[33] T. Hida and Si Si, Innovation for random fields, Infinite Dimensional Analysis, Quantum Probability and Related Topics. 1 (1998), 499–509.

[34] T. Hida and Si Si, Stochastic processes with Poisson noise innovation, Quantum Information and Complexity, Part II, World Scientific Pub. Co.

[35] T. Hida, Si Si and Win Win Htay, Variational calculus for random fields parametrized by a curve or a surface, to appear.

[36] T. Hida, White noise analysis: A new frontier, Volterra Center Pub., N. 499, January 2002.
[37] T. Hida, White noise and functional analysis, Seminar on Prob. (in Japanese) 2002.
[38] M. Hitsuda, Multiplicity of some class of Gaussian processes, Nagoya Math. J. 52 (1973), 39–46.
[39] E. Hopf, Statistical hydrodynamics and functional calculus, J. of Rational Mechanics and Analysis. 1 (1952), 87–123.
[40] A. Jaffe, A. Lesniewski and K. Osterwalder, Quantum K-theory, Commun. Math. Phys. 118 (1998), 1–14.
[41] John R. Klauder and E.C.G. Sudarshan, Fundamentals of Quantum Optics, Math. Phys. Monograph series W. A Benjamin, Inc. 1968.
[42] H.-H. Kuo, White noise distribution theory, CRC Press. Probability and Stochastics Series, 1996.
[43] R. Léandre and H. Ouerdiane, Comes-Hida calculus and Bismut-Quillen super connections, Les prépublications de l'institut Elie Cartan. 2003/no. 17.
[44] Y.J. Lee, Generalized functional of Poisson noise.
[45] P. Lévy, Théorie de l'addition des variables aléatoires, Gauthier-Villars. 1937.
[46] P. Lévy, Processus stochastiques et mouvement browniwn. Gauthier-Villars, 1948. 2ème ed. 1965.
[47] P. Lévy, Problèmes concrets d'analyse fonctionnelle, Gauthier-Villars, 1951.
[48] P. Lévy, Random functions: General theory with special reference to Laplacian random functions, Univ. of California Publ. in Statistics vol. 1. (1953), 331–390.
[49] P. Lévy, Random functions: A Laplacian random function depending on a point of Hilbert space, Univ. California Publ. in Statistics vol. 2. (1956), 195–206.
[50] P. Lévy, A special problem of Brownian motions and a general theory of Gaussian random functions, Proc. 3rd Berkeley Symp. on Math. Stat. and Probability vol. II (1956), 133–175.
[51] P. Lévy, Fonctions aléatoires à corrélation linéaires, Illinois J. of Math. 1 (1957), 217–258.
[52] P. Lévy, Jacques Hadamard, sa vie et son oeuvre – Calcul fonctionnel et questions diverses, La vie et l'Oeuvre de Jaque Hadamard. Monographie N0. 16 de l'Enseignement Mathématique. Imprimerie Kundig, Genève, (1967), 1–24.
[53] P. Lévy, Foctions de lignes et équations aux dérivées fonctionnelles, Comm. XIII Congréss international d'histoire des Sciences, Moscow, 1971.
[54] J.L. Lions and E. Magenes, Non-homogeneous boundary value problems and applications, vol. I, Springer-Verlag, 1972.
[55] R.S. Liptzer and A.N. Shijarev, Statistics of random processes, Springer-Verlag, 1977.
[56] H.P. McKean, Brownian motion with a several-dimensional time, Theory of Prob. Appl. 8 (1963), 335–354.
[57] H.P. McKean, Stochastic integrals, Academic Press, 1969.
[58] A.S. Monin and A.M. Yaglom, Statistical Hydrodynamics, Nauka, vol. 1, 1965; vol. 2, 1967. (In Russian)

[59] E.A. Morozova and N.N. Chentsov, P. Lévy's random fields, Theory of Prob. Appl. (1949) 153–156.
[60] E. Nelson, Construction of quantum fields from Markov fields, J. Functional Analysis 12 (1973), 97–112.
[61] E. Nelson, The free Markov field, J. Functional Analysis 12 (1973), 211–227.
[62] N. Obata, A characterization of the Lévy Laplacian in terms of infinitedimensional rotation groups. Nagoya Math. J. 118 (1990), 111–132.
[63] N. Obata, White noise calculus and Fock space. Lecture Notes in Math. #1577, Springer-Verlag, 1994.
[64] Y. Okabe, On a stationary Gaussian process with T-positivity and its associated Langevin equation and S-matrix. J. faculty of Sci., University of Tokyo, Sec. IA 26 (1979), 115–165.
[65] Pontrjagin, Optimal Control.
[66] H. Poincaré, Les méthodes nouvelles de la mécanique céleste, Tome III, Gauthier-Villars, 1899.
[67] R.T. Rockafellar and R. Wets, Variational analysis, Springer-Verlag, 1998.
[68] K. Saito and H. Tsoi, The Lévy Laplacian acting on Poisson noise functionals. Infinite Dim. Analysis, Quantum Prob. and Related Topics 2 (1999), 503–510.
[69] K. Saito, A stochastic process generated by the Lévy Laplacian. Acta Applicandae Math. 63 (2000), 363–373.
[70] Si Si, A note on Lévy's Brownian motion I,II, Nagoya Math. J. vol. 108 (1987), 121–130, vol. 114 (1989), 165–172.
[71] Si Si, Variational calculus for Lévy's Brownian motion "Gaussian random fields" Series Prob. & Statistics vol. 1, World Sci. Pubilications (1991), 364–373.
[72] Si Si, Integrability condition for stochastic variational equation, Volterra Center Pub., Univ. di Roma Tor Vergata vol. 217, 1995.
[73] Si Si, Innovation of some random fields. J. Korean Math. Soc. 35, 1998, No. 3, 575–581.
[74] Si Si, A variation formula for some random fields; an analogy of Ito's formula Infinite Dimensional Analysis, Quantun Probability and Related Topics, vol. 2. World Scientific.
[75] Si Si, Topics on random fields, Quantum Information I, eds. T. Hida and K. Saito, World Scientific.
[76] Si Si, White noise approach to random fields, to appear in the proceeding of International summer school, "White noise approach to Classical and Quantum Stochastic Calculus".
[77] Si Si, Gaussian processes and Gaussian random fields, Quantum Information II, eds. T. Hida and K. Saito. 2000 World Scientific, 195–204.
[78] Si Si, Random fields and multiple Markov properties, Supplementary Papers For the 2nd International conference on Unconventional Models of Computation, UMC'2KI. eds. Antoniou *et al.*, CDMTCS-147 Center for Discrete Mathematics and Theoretical Computer Science, Solvay Institute 2001, 64–70.
[79] Si Si and Win Win Htay, Entropy in subordination and Filtering, Recent developments in infinite dimenional analysis and Quantum probability, Accardi *et al.*

[80] Si Si, Note on Poisson noise, to appear in "Quantum Information and Complexity, Part II", World Scientific Pub. Co.
[81] Si Si, Effective definition of Poisson noise, Infinite Dimensional Analysis, Quantum Probability and Related Topics. to appear.
[82] B. Simon, The $P(\phi)_2$ Euclidean (Quantum) field theory. 1974. Princeton Univ. Press.
[83] L. Streit and T. Hida, Generalized Brownian functionals and the Feynman integral, Stochastic Processes and their Applications. 16 (1983), 55–69.
[84] L.T. Todorov, M.C. Mintchev and V.B. Petkova, Conformal invariance in quantum field theory, Scuola Normale Superiore Pisa. 1978.
[85] L. Tonelli, Fondamenti di Calcolo delle variazioni, Bologna Nicola Zanichelli Editore. vol. 1, 1921; vol. 2, 1923.
[86] V. Volterra, Leçons sur les fonctions de lines, Gauthier-Villars, 1913.
[87] V. Volterra, Le calcul des variations, son évolution et ces progrés son rôle dans la Physique mathématique. Praha-Brno. 1932.
[88] V. Violterra and J. Pérès, Théorie générale des fonctionnelles, Gauthier-Villars, 1936.
[89] V. Volterra, Theory of functionals and of integral and integro-differential equations, Dover, 1959.
[90] V. Volterra, Oevre V. (1881–1892).
[91] N. Wiener, The homogeneous chaos, Amer. J. of Math. 60 (1938), 897–936.
[92] N. Wiener, The discrete chaos, Amer. J. of Math. 65 (1943), 279–298.
[93] N. Wiener, Nonlinear problems in random theory, Technology Press of M.I.T. 1958.
[94] Win Win Htay, Linear process; some new topics, Proc. of Quantum Information and Complexity, Meijo Univ. 2003. World Scientific Pub. Co. (to appear).
[95] Y.A.M. Yaglom, Some classes of random fields in n-dimensional space, related to stationary random processes. Theory of Prob. Appl. 2. (1957) 273–319, summary 319–320.
[96] K. Yosida, Functional analysis, 6th ed. Springer-Verlag, 1980.
[97] Broken symmetry, Selected papers of Y. Nambu, World Scientific Pub. Co. Ltd, 1995.

Index

Printed in the United States
By Bookmasters